VOX EX MACHINA

VOX EX MACHINA

A Cultural History of Talking Machines

SARAH A. BELL

The MIT Press
Cambridge, Massachusetts
London, England

The MIT Press would like to thank the anonymous peer reviewers who provided comments on drafts of this book. The generous work of academic experts is essential for establishing the authority and quality of our publications. We acknowledge with gratitude the contributions of these otherwise uncredited readers.

This book was set in Adobe Garamond and Berthold Akzidenz Grotesk by Westchester Publishing Services. Printed and bound in the United States of America.

Library of Congress Cataloging-in-Publication Data is available.

ISBN: 978-0-262-54635-5

10 9 8 7 6 5 4 3 2 1

One day ladies will be walking their computers in the park and say, "Do you know; my little computer said a very funny thing to me this morning!"
–Alan Turing

To live effectively is to live with adequate information.
–Norbert Wiener

The human voice is the most beautiful instrument of all, but it is the most difficult to play.
–Richard Strauss

Contents

Acknowledgments

My interest in voice synthesis began in 2011—not in October, when Apple announced Siri integration for the iPhone 4S, but earlier that year, over the Fourth of July weekend, when my two teenaged children had me drive them and some of their friends, in a rented minivan, from the Salt Lake Valley across the desert to Los Angeles, so they could attend a concert at Anime Expo given by Hatsune Miku, a virtual singer whose voice is generated with voice synthesis software. The curiosity and creativity of my kids have always inspired me, and I am so proud of the adults they have become. I'm lucky to be their mom.

I am grateful for the expertise and assistance of the archivists, librarians, and special collections staff at the libraries and museums that I visited while conducting research for this book: the DeGolyer Library at Southern Methodist University; the Brian Sutton-Smith Library & Archives of Play at the Strong Museum of Play; the Computer History Museum; the Archives Center at the Smithsonian National Museum of American History; the Library of Congress; Distinctive Collections at the MIT Libraries; and the Special Collections at the Green Library at Stanford University. Much of this research was funded by the Association for Computing Machinery History Committee, the Strong Research Fellowship from the Strong Museum of Play, a Lemelson Center Travel Award from the Lemelson Center for the Study of Invention and Innovation, and a Scholarship and Creativity Grant from the Research Excellence Fund at Michigan Technological University.

I was extremely fortunate to be selected as a Digital Studies Fellow at the John W. Kluge Center at the Library of Congress, where I spent most of

2018 completing the research for this project. I am thankful for the support and friendship of the Kluge program manager Travis Hensley, program assistants Emily Coccia and Michael Stratmoen, intern Catherine Able Thomas, and fellow resident scholars Katie Booth, Andrew Meade McGee, Seonaid Rogers, and Kristen Shedd.

This project started as my doctoral thesis and had the enthusiastic support of my trusted advisor, Maureen Mathison, and committee members Robert Gehl, Sean Lawson, and especially Glen Feighery. Glen, I'm sorry that "O, Superman!" didn't make it into this final version! Other University of Utah Communication faculty who prepared me to be able to see this project through include Helene Shugart, Kevin DeLuca, Kent Ono, Connie Boros, April Kedrowicz, and Natalie Stillman-Webb. At Michigan Tech, I am grateful for the friendship and support of my colleagues, especially the "media club" of Erin Smith, Abraham Romney, Karla Kitalong, Diane Shoos, and Kette Thomas. I was fortunate to be nurtured as an undergraduate English major by Carol A. Martin, a remarkable scholar of Victorian literature whose enthusiasm for "fat books by British women" showed me that scholarship can be a joy as well as an intriguing mystery, and who picked me up at a low time in my life and deposited me into graduate school.

I am beyond thankful for the patience and good humor of the acquisitions editor Katie Helke and the staff at the MIT Press. Early advice from Christopher Lura Editorial was most helpful as well. The community fostered within the SHOT Special Interest Group for Computers and Information in Society has been a welcome source of advice, encouragement, and collaboration and provided early opportunities to present and get feedback on the case studies in this book.

Finally, this book is dedicated to my mother, Janet M. Davis. For everything.

PROLOGUE

In the spring of 1770, as Maria Theresa, queen of Bohemia, Hungary, and Croatia, was marrying off her fifteenth child, Marie Antoinette, to the future King Louis XVI of France, a trusted civil servant named Wolfgang von Kempelen presented Her Majesty with a demonstration of what would become one of the most famous mechanical automatons of all time—*den Schachspieler von Kempelen*—the Chess Player of von Kempelen, nicknamed "the Turk" because it was dressed in an ermine-trimmed robe, loose trousers, and turban in the Ottoman style and held a long Turkish pipe in its left hand.[1]

The figure itself was carved of wood and sat behind a large cabinet on the top of which was displayed a chessboard. In the demonstration, von Kempelen made a show for the queen, unlocking one of three doors on the front of the cabinet to reveal an intricate mechanism of wheels, gears, and other moving parts for the audience to scrutinize. Like a magician proving to the audience that there is no trick, von Kempelen opened the other doors to show that there was no one hiding in the cabinet. He spun the apparatus around and exposed the gearwork in the automaton itself. He made an elaborate show of setting up the chess pieces and preparing the Turk for the game, winding up the machine with a large key. A courtier volunteered to play against the machine, which would shake its head if its human opponent tried to cheat. The *Schachspieler* didn't speak, but, amazingly, could spell out on a letterboard answers to questions posed to it from the audience. The automaton beat its first human challenger handily, delighting the queen and ensuring that the spectacle would be performed over and over again to entertain her guests. Word of the mechanical marvel spread, and von Kempelen was entreated to take the *Schachspieler* on tour. But von Kempelen, who came to refer to his creation derisively as a "bagatelle," would have rather spent his time focused on his real passion, inventing mechanical speech.[2]

As science historian Jessica Riskin has shown, the Chess Player's medieval antecedents likely influenced René Descartes's concept of the animal-machine, the idea that bodies, including human bodies, were essentially machinery and could be understood as such. Although this radical idea was sometimes mocked in Descartes's own time, Riskin contends that the animal-machine represents an epistemological revolution that changed how people went about understanding and explaining natural phenomena.[3] Eighteenth-century automata bear this out. As playthings of the aristocracy, automata became even more elaborate, expensive, and exquisite. Craftspeople developed techniques to manufacture the tiniest precision components, which designers used to create ornate and ingenious imitations of nature—glass that looked like flowing water, silver scales that moved like feathers on a swan's neck, or porcelain fingers holding a quill pen or playing a flute. Intricate cams allowed designers to program their automatons to perform sequential movements in imitation of living things themselves—writing by hand, pressing keys on a musical instrument, or dancing a waltz. Even beyond this, designers of automata sometimes fashioned their creations to mimic the most biological acts of the body: breathing, eating, bleeding, and even defecating. Many believed that the body was as knowable as a machine, and scatological automata were the proof.

Debates ensued about the nature of human life itself. Within this environment, von Kempelen's Chess Player challenged assumptions about whether mental processes could be produced from artificial machinery.[4] Much later, it would come to light that the Turk did not actually play chess on its own. Meanwhile, von Kempelen, the perpetrator of the ruse, who knew that his automaton was only capable of moving pieces around a magnetized board at the command of a human chess master hidden inside, was interested in solving the seemingly more achievable problem of giving a mechanical person, like the Turk, a voice. By 1783, von Kempelen had developed a mechanism that could produce the vowel and most of the consonant sounds of his native German at a squawky monotone, although it took quite a bit of skill to manipulate his voice machine to combine these sounds into recognizable words and phrases. His simple analog of the body's vocal system was created from a bellows that blew air across the drone reed from a bagpipe and through a rubber funnel "mouth." The operator used one hand to open up or close off the "mouth" while his elbow pressed down on the bellows "lungs" to create the "breath." Not surprisingly, the results sounded nothing like the human voice, but it might be recognizable as speech with some effort.[5] In 1791, von Kempelen published a detailed description of this speaking machine and his study of the human voice on which its construction was based.[6]

In spite of his deception with the Chess Player, von Kempelen's speech research was accepted as scientific. The eminent Victorian physicist Sir David Brewster declared in 1837, "We have no doubt that, before another century is completed, a talking and a singing machine will be numbered among the conquests of science."[7] Brewster was off by almost 100 years. It would be near the end of the twentieth century before a machine-generated voice approached sounding human.

The key to achieving that aim would prove to be electricity. While von Kempelen was experimenting with his bellows, an Italian contemporary, Luigi Galvani, was documenting experiments in bioelectricity. Galvani published a book of his research in the same year as von Kempelen's, asserting ideas about an electrical power that animated living things, which he called "animal electricity." One of Galvani's countrymen, Alessandro Volta, who disagreed with him about the source of this electric vitalism, was conducting experiments of his own that would result in the first electrochemical battery in 1799, a means of generating electrical charge that set the stage for the nineteenth century's developments in harnessing electricity for human use. However, the idea of "animal electricity" took hold in the imagination of other scientists, philosophers, and poets. One believer was Erasmus Darwin, a physician and grandfather of Charles, who had a keen interest in the artificial production of life. He also experimented with creating mechanical voices. In a footnote to his poem *The Temple of Nature* (1802), Darwin seems to have constructed a variation of von Kempelen's machine: "I contrived a wooden mouth with lips of soft leather, and with a vale back part of it for nostrils, both which could be quickly opened or closed by the pressure of the fingers, the vocality was given by a silk ribbon . . . stretched between two bits of smooth wood a little hollowed; so that when a gentle current of air . . . was blown . . . it gave an agreeable tone . . . much like a human voice."[8]

The application of electricity to voice synthesis was several decades in the future, but the imagination for electricity as a life force was part of the zeitgeist of early nineteenth-century European art and science. One evening in 1816, a group of young poets sat around a Swiss villa discussing the experiments of Erasmus Darwin and others. Later that night, one of them, eighteen-year-old Mary Shelley, dreamed about a corpse reanimated by electricity and woke to start writing *Frankenstein*.[9]

These purely mechanical attempts to synthesize the human voice weren't very successful, but studies of the voice and human vocal system often underlaid other developments in electrical communication. William Cooke and Charles Wheatstone gave the first commercial demonstration of an electric telegraph in London in 1837. Wheatstone's research interest was acoustics, experimenting with the movement of sound waves through various materials. One of these experiments was meant to improve on von Kempelen's speaking machine. Wheatstone's machine inspired the teenaged Alexander Graham Bell, whose father took him to meet Wheatstone, an experience that Bell said led him "onward irresistibly" toward developing the telephone he would eventually patent.[10] "Although the articulation was disappointingly crude, it made a great impression upon my mind," recalled Bell about Wheatstone's talking machine.[11] Wheatstone loaned the Bells his copy of von Kempelen's book, and young Alexander "devoured it."[12]

He and his brother set out to create their own talking automaton, with their father challenging them to try to more closely imitate nature rather than copying von Kempelen's model. Alexander focused on the mouth and tongue, casting them from a real human skull using the Victorian miracle plastic gutta-percha, while his brother, handier with tools, took on the larynx and vocal folds.[13] Bell described quite a contraption of levers and rubber and wood, complete with an artificial larynx constructed from tin, and a flexible tube windpipe that could "squeak like a Punch and Judy show."[14]

When the throat and mouth were put together, even before the wood and rubber tongue was in place, the boys "obtained a quite startling reproduction of the word 'mamma' pronounced in the British fashion," which offered an excellent opportunity to play tricks on the neighbors.[15] Bell recalled that the speaking head didn't progress far beyond that point, but it had realized his father's desire "that through its means his boys should become thoroughly familiar with the actual instrument of speech and the functions of the various vocal organs," experiments which took advantage of an unfortunate stray cat, a lamb's throat supplied by the local butcher, and the family's very patient terrier manipulated to growl out the sounds "Ow-ah-oo-gamama" ("How are you, grandmamma?" if you "exercise a little imagination").[16]

As the animal-machine had been a key paradigm for understanding the body in the eighteenth century, the electrical system became the key analogy of embodiment for some scientists in the nineteenth. Along with the development of the electric telegraph, a metaphor developed through which electricity became the carrier of communication, both within and without the human body. Even if electricity could not bring to life a Frankenstein monster, it could bring to life, "like the great brain," instantaneous communication through the telegraph, or "the nervous system of Britain," as the Victorian physician Andrew Wynter put it in 1854.[17] As for explaining what happened inside the body, the future Royal Society president William Crookes speculated in 1892 that electrical organs inside human brains might allow thoughts to be transmitted telepathically through an electromagnetic ether. The German polymath Hermann von Helmholtz thought of the human nervous system as a telegraph communicating sense perceptions through the body as electrical information.[18]

This paradigm would be refined through twentieth-century cybernetics, which conceived of information as external to the body, something in the environment that is processed in the human nervous system and is exemplified in the cybernetic metaphor of the electronic stored-program computer as a "brain." A subtle but significant shift follows: von Helmholtz understood sound as sensation. That is, he was not focused on the *source* of sounds, how they are *produced*, but only on what happens when sound waves are *heard*. The sound studies scholar Jonathan Sterne summarized this shift in *The*

Audible Past, his study of nineteenth-century sound-recording technologies: "The main advancements in sound research and in sound reproduction came from abandoning the mouth altogether."[19] This plays out as Sterne describes for Alexander Graham Bell with the telephone and Thomas Edison with the phonograph, their inventions situated within a trajectory of research about the ear. There is no doubt that the scientific and cultural origins of sound recording interacted as Sterne describes—that the phonograph and the telephone changed what it meant to listen, how people listened, what they listened to, and why. As Sterne later documented in *MP3: The Meaning of a Format*, this focus on what is heard continues into the digital age as sound file compression algorithms are designed to eliminate data representing parts of the audio waveform that are less likely to be heard, in a process called perceptual coding.[20] Sound engineering on the whole is focused thus; however, Bell's namesake company, and, later, its research and development arm, the Bell Telephone Laboratories, also invested in *speech* research, and these scientists and engineers, far from abandoning study of the human vocal system, would dig in.

INTRODUCTION

In November 2022, Amazon announced significant layoffs within the division that develops its Alexa voice assistant products. The business press reported that Alexa, a pet project of Amazon founder Jeff Bezos, who imagined his company creating a real-world version of the talking computer from the *Enterprise* spaceship on *Star Trek*, has never been profitable.[1] True to form, Amazon sold its Alexa-enabled speakers at near cost, expecting to see profits when customers used the voice assistant to make other Amazon purchases, subscribe to premium streaming audio services, and use third-party applications, which it calls Alexa Skills (like ordering a pizza from a restaurant, for which Amazon might take a percentage). Instead, customers seemed mostly to use them as DJs, kitchen timers, and to ask about the weather. Former employees reported the division had been "in crisis" since at least 2019, and Google and Apple have also struggled to see profits from their own voice assistant applications.[2] In spite of a spate of articles predicting that voice assistants were "doomed," the release of ChatGPT on November 30, 2022, quickly revived predictions that voice, when combined with generative text artificial intelligence (AI), will be the information interface of the future.[3]

Such predictions long precede Alexa. If the telegraph marks the beginning of electronic information and communication technologies (ICTs), then the idea of improving ICTs by enabling human-machine voice communication has been an engineering goal from the beginning.[4] As discussed in the prologue to this book, telegraph pioneer Charles Wheatstone experimented with mechanical voice synthesis, inspiring the young Alexander

Graham Bell. That the public was aware of experiments like Wheatstone's is evidenced by a satire published in the popular nineteenth-century British magazine *Punch*, attributed to the Victorian writer William Makepeace Thackeray, suggesting that voice synthesis could be added to Charles Babbage's Analytical Engine mechanical calculator or even a printing press.[5] More realistically, the binary encoding of language in Morse code's dits and dahs provided a model for voice coding developed by early twentieth-century telephone engineers looking to improve the efficiency of their growing networks. These networks, widely described at the time as society's "nervous system," were soon to be managed by the "electronic brains" of stored-program computers, and it logically followed that electronic brains could use electronic voices to communicate. Voice synthesis for computers emerged in the 1960s, 1970s, and 1980s, though it was significantly buzzy and electronic sounding by today's standards. Corporations including AT&T, IBM, Texas Instruments (TI), and Apple, as well as dozens of smaller companies and research labs, developed and improved voice synthesis throughout the twentieth century, following the seemingly common-sense idea that the easiest way for humans to interact with machines was simply to have them talk to us.

This book follows the engineering quest to simulate the human voice using electronic machines across the twentieth century. But it challenges a truism often cited by voice technology developers who reason that voice is the most natural way for people to interact with machines, the easiest way, and the most desirable. Instead, I propose that conversing with machines is not *natural* at all, but cultural, learned, and negotiated. Talking machines only seem natural because we have been primed, both by digital technology companies and popular media, to expect to interact with machines by voice. In fact, listening to machines is often not easy, never mind talking with them. Recent attention to many negative consequences of big tech—racist algorithms, social media manipulation, misinformation, disinformation, and surveillance, among others—should at the very least give us pause to consider the implications of adopting another unregulated technology, especially one that simulates the human voice.[6] After all, voice is the main modality through which we negotiate our social world with other people. As computational and networked smart devices are embedded further into our daily lives, the fact of their having human-sounding voices raises important questions about how their use may manipulate our interactions with the corporations that develop and deploy them, and especially their impact on our interactions with other human beings. Voice interface applications have reached the point where their widespread deployment is practical, making it critical that we ask what is at stake in their use, especially as two of the most rapidly developing service areas for voicebot interaction are mental health care and assistive care, both needed by people when they might be particularly vulnerable.

My purpose in writing this book is to give people, those of us that the tech industry calls "users," some understanding of voice synthesis as a *product* in order to help us make informed decisions, individually and collectively, about the role that we will allow it to play in our daily lives. I have chosen the word "product" deliberately to imply more than just an interface, more than just another feature of another gadget crowding our backpacks and purses, our commutes and cars, and our homes and bedrooms. A product is not a social companion. Voicebots do not experience communication as people do, an obvious fact that nevertheless gets lost in Silicon Valley hype, as well as our own tendency to accommodate them. As a simulation of one of our most potent means of human expression, voice synthesis reveals the limitations of computational technologies when compared to our biological abilities, as well as for fulfilling our needs as human beings. We are more than just users.

In telling this history, I look at the *development* of voice synthesis technologies to reveal the assumptions about the human voice that synthesis is modeled on, including its limitations, and the differences between synthesized voices and the voices that our bodies can produce. I also look at the *deployment* of voice synthesis technologies by various US corporations to reveal its use as a choice among options, motivated by corporate values, especially efficiencies that often serve the corporation at the expense of people. Finally, I look at some *public receptions* of voice synthesis by analyzing media coverage of specific voice synthesis products when they were introduced. Consumers have often rejected the limitations of voice synthesis and have questioned the role of talking machines in human lives. Reflecting on these previous debates can inform our own. Each chapter attempts this three-part discussion of a specific voice synthesis product. The chapters are arranged in chronological order to highlight both continuities and disruptions in the technical development of, and cultural ideas about, voice synthesis across the twentieth century.

The earliest experiments in electronic voice synthesis were created at AT&T's Bell Telephone Laboratories (colloquially referred to as Bell Labs) in the 1920s and 1930s as part of that company's quest for worldwide, universal telephone service. This was during the Machine Age, the early twentieth-century period between the world wars during which electrically powered and controlled machinery was rapidly being developed for industrial and home use and for media communication, sparking public discussions about automation, consumerism, and propaganda that would become endemic to the Information Age. In fact, electronic voice synthesis indexes the Information Age because it is an information technology that made its public debut in 1939, just ahead of the electronic computers that enabled the later twentieth-century information-based economy that is the hallmark of our era.[7] Voice synthesis developed alongside computational technologies that were introduced into the popular imagination as "electronic brains" and described from their beginning as "speaking," an implicit part of the "brain" metaphor. Experience with voice synthesis provided a way for the US public to negotiate what the adoption of computational technologies meant for social, cultural, and economic life as it followed the spread of computing from military and industrial applications, into business, education, and the home.

Bell Labs research about the human body's vocal system contributed to knowledge about signal processing that furthered AT&T's goals for telecommunication, even as actual voice synthesis found a different commercial application as an interface for human-computer interaction (HCI). Throughout most of this development, the goal of researchers and engineers was to develop intelligible voice synthesis. The choice of the term "intelligible" is important because it reveals a general belief in Alan Turing's argument that if two information processors (say, a human brain and a computer) produce indistinguishable output, then they are essentially equivalent regardless of their internal operations.[8] Voice synthesis was enrolled in the AI agenda as a warrant for the argument that the manipulation of symbolic language is the cornerstone of intelligence, even if voice—giving sound to that language—was relegated to the lower status of mere medium.

Even before the statistical models of natural-language processing (NLP), data accumulation, and computer processing power that made speech recognition a reality were available, the conceptual foundation of computational machines as independently intelligent by virtue of their ability to communicate was well established. Voice synthesis was a potent public proof of concept, and one whose limitations allowed for the social and cultural negotiation of what human being means in relation to computational machines that might imitate processes and behaviors once thought to be uniquely human—like speech.

This book offers a history of electronic voice synthesis that shows how it has achieved a modicum of success as a simulation of the articulation of human language but is fundamentally limited in its ability to simulate expressive human communication. This is generally a good thing because it helps us differentiate between the voices of computer applications and the voices of other people, an important distinction for avoiding bad actors who would use this technology to deceive us. But the downside is that habituation to daily life with potentially less vocal diversity, due to the pervasive deployment of synthesized voices, might make us less prepared to effectively negotiate our social interactions with other people. This is at the heart of concerns about children learning that it is OK to use bad manners or abusive speech with voice assistants, and about how negative gender stereotypes may be perpetuated by exposure to female-sounding voice assistants, but these specific issues are rather easily dealt with. The deeper problem remains, however, that less exposure to human vocal diversity leaves us less prepared to understand, value, and benefit from our interactions with other people. The answer isn't a proliferation of experiences with a greater diversity of synthesized voice simulations, but rather more frequent exposure to the expressive vocal diversity of human beings. This book shows why there are limits to the prosocial benefits of voice synthesis and highlights some of the risks associated with its deployment.

THE INFORMATIC VOICE

Voice synthesis has become sophisticated enough now that it is easy to ignore the fact of its simulation. One point of this book is to draw attention to that fact. To do so, I highlight the nature of voice synthesis as *informatic*. The *Oxford English Dictionary* defines "informatic" as relating to informatics *or* information, but I use it here to stress that voice synthesis is a product of informatics *and* information.[9]

Information has an interesting etymology, beginning with the ancient scholastic tradition where *to inform* related to the formation of the mind and character by knowledge. Later, the word "information" came to stand for knowledge itself, closer to our contemporary colloquial usage, as in, "I'd like some information about that." In the early twentieth century, a new, technical definition developed, with "information" coming to mean, within communications engineering, "a mathematically defined quantity divorced from any concept of news or meaning; especially one which represents the degree of choice exercised in the selection or formation of one particular symbol, message, etc., out of a number of possible ones, and which is defined logarithmically in terms of the statistical probabilities of occurrence of the symbol or the elements of the message."[10] Language is represented by symbols (e.g., letters), and it was discovered that there were underlying statistical patterns to the arrangement of those symbols that could be separated from what the symbols were meant to convey in order to transmit language more quickly and accurately. In other words, form could be separated from meaning.

This abstraction is fundamental to almost everything that we can accomplish computationally. Computers don't understand language; they simply process it mathematically. They can process so much of it, so quickly (and according to statistical properties that many of us don't entirely understand), that it can seem like they understand it. A second key to this illusion is the conversational paradigm that has been used to describe people's interactions with computers since the very first mainframes were invented. Use of the voice as a synecdoche to stand in for communication already existed— we refer to political participation as "having a voice"; we "hear" an author's "voice" in their writing; we talk of people "claiming a voice" when they assert their agency. And many people are used to the idea of a metaphorical, omniscient, and disembodied voice of God or multiple gods.[11] Similarly, the output provided by computers was consistently referred to as "communication" delivered by the computer speaking. Electronically produced simulation of the human voice precedes the invention of electronic stored-program computers, so even as early computers were described as having metaphorical voices, research was underway to enable them to talk, literally. The means of doing this was to identify statistical patterns in the waveforms of voice signals—in other words, to create voices *out of information*. At the same time, some computer developers thought that computer voices were needed to deliver computer output, another sense of how we use the word "information," in a format supposedly easy for any person to understand.

"Informatic" draws attention to some additional aspects of voice synthesis that help us think about it in contrast to the human voice that it simulates. "Informatics" refers to the post–World War II discipline that deals with the structure, properties, and communication of information and with the means of storing or processing information.[12] In the second half of the twentieth century, this was increasingly accomplished with computers and is central to shifts in society, politics, and the global economy that distinguish this period in history, continuing into our own, as the Information Age. One of these shifts has been the mass digitization of physical media (e.g., Google's project to scan all the world's books), resulting in faster, easier, and cheaper manipulation and circulation of the content stored in those formats. The body has also become an information-carrying medium in the Information Age. We speak of our deoxyribonucleic acid (DNA) as a genetic code that contains all the information about our physical bodies. We can track our own biometrics using digital sensors and understand things about ourselves in terms of the information that those sensors can collect and process. And voice synthesis is a digital simulation of what happens in the vocal system of a human body.

Although modeled on the human vocal apparatus, the informatic voice is a product of electronic signal processing. The word "informatic" highlights the fact that voice synthesis is an abstract mathematical simulation of the biophysical process in a human body. The individual, dynamic, and sometimes uncontrollable assemblage of flesh, blood, bones, hormones, cartilage, muscles, and sinews that make up the human throat, larynx, mouth, tongue, and lips was simplified, normalized, measured, modeled, and abstracted in electronic circuits and filters representing only the most basic characteristics of that bodily messiness: pitch and phonemic articulation. As computer-processing speed and memory increased, these synthesis models increased in detail, but the informatic voice retains sonic artifacts of being *processed information* rather than being biologically produced. This is why it is so easy for us to poke fun at the lilting cadence and limited emotion of an informatic voice like Siri. The latest voice synthesis technologies promise to respond to us with increased emotion in their voices, but computers don't experience emotion; they only calculate. The illusion of their vocal authenticity relies on our willingness to accept the informatic model that turns the human body itself into an information-processing system and defines communication only in terms of information exchange.

"Informatic" has been used in media theory similarly to how I use it here, often to critique the way that the digitization of images was affecting society at the end of the twentieth century, when software-driven "new media" became pervasive. At one extreme, the semiotician Marshall Blonsky mourned the culture in 1992, stating, "The system *we* inhabit destroys the taste for origins, referents, reasons; *they have been* devoured by the image. And of course *we have become less* and *less physical. Everything becomes informatic, everything becomes soft.*"[13] Others seemed to welcome the new possibilities that digitization offered, including artist Lev Manovich, who called the relational database for organizing information "a new symbolic form of the computer age."[14]

The body was also imbricated in the informatic nature of the digital. The new media scholar José van Dijk, in her 2007 study of digital imagery's cultural impact, explained about imaging the brain, "When digitizing the body, the biological is rearticulated as informatic in order to be enhanced or redesigned. Both body and machine are considered platforms through which activities are mediated, yet the materiality of that platform profoundly *matters*: information is embodied as much as flesh is computed."[15] Stress on the word "matters" was meant to draw attention to its dual meanings of importance ("it matters") and physicality ("made of matter"). She did not believe that human memory, for example, was exclusively informatic, but rather a phenomenon of the specific material of the human brain, and could not be separated from it; memory could not be reconstructed from brain imaging in the way that an image itself can be reconstructed from the numeric data that represents that image in its digital form. Over the last thirty years, new media scholars have debated the implications of this kind of abstraction of the body when other human attributes and abilities have been targets of digitization.

For example, in 1999's *How We Became Posthuman*, literary theorist N. Katherine Hayles questioned the emerging transhumanist idea that human consciousness could be stored as information, and she found the technological fantasy of the complete digitization of a human mind impossible and the dream of eternal digital life misguided. She recognized that "a defining characteristic of the present cultural moment is the belief that information can circulate unchanged among different material substrates," and quipped that "it is not for nothing that 'Beam me up, Scotty,' has become a cultural icon for the global informational society."[16] In looking at the history of cybernetics and later visions that some technologists expressed for transcending the "meat sack" of the human body, Hayles defined a particular orientation to the informatic that she termed "posthuman" as "privileg[ing] informational pattern over material instantiation, so that embodiment in a biological substrate is seen as an accident of history rather than an inevitability of life." She identified several assumptions that this posthuman orientation depended on:

> The posthuman view considers consciousness, regarded as the seat of human identity in the Western tradition long before Descartes thought he was a mind thinking, as an epiphenomenon, as an evolutionary upstart trying to claim that it is the whole show when in actuality it is only a minor sideshow. . . . [T]he posthuman view thinks of the body as the original prosthesis we all learn to manipulate, so that extending or replacing the body with other prostheses becomes a continuation of a process that began before we were born. . . . . [B]y these and other means, the posthuman view configures human being so that it can be seamlessly articulated with intelligent machines. In the posthuman, there are no essential differences or absolute demarcations between bodily existence and

computer simulation, cybernetic mechanism and biological organism, robot teleology and human goals.[17]

This posthuman perspective replaces both religion and philosophy in providing a comprehensive account of the past, present, and future of human beings as migrating information that can be reembodied, perhaps infinitely. Even as she found the assumptions of the posthumanist view problematic, Hayles was interested in the impact of this posthuman subjectivity as a prevailing ideology.

Almost a quarter of the way through the twenty-first century now, and in the midst of several existential crises, from climate change to expanding authoritarianism, my interest in calling attention to the informatic nature of voice synthesis is more explicitly critical of the primary posthuman assumption that privileges informational pattern over material instantiation, and of the power of its concomitant economic ideology over the last seventy-five years. Because it has been so successful, I find it imperative to assert that there *are* essential differences and absolute demarcations between bodily existence and computer simulation. Histories of technology, especially those that attend to the political and economic contexts of technology development, help show who benefits most from the informatic imperatives of our age.

Attending more closely to the way that communication itself became informatic, the media theorist Bernard Dionysius Geoghegan has argued that information theory (which defined the statistical properties of language) "set in motion a new scientific strategy according to which the natural, cultural, and technical worlds could be described in terms of statistically patterned data flows," where everything from genetics to psychological disorder was seen as a process of *communication*.[18] As we see today, the ability to control data flows provides significant political, economic, and social power. As Geoghegan observed, with human beings defined by this logic as the channels, rather than the authors, of data streams, those who control the infrastructure for those data streams get to control the data.

With communication redefined as the efficient circulation of statistically patterned data flows (and almost always for economic advantage), there is no place in this system for communication as a social act. Meta chief executive officer Mark Zuckerberg has referred to interpersonal interactions on

Facebook as "social mechanics" because the goal of the site is to constrain communication so that it is easily "mined" as data that can be used to profile users for its advertisers.[19] For example, Facebook entices users to express their emotional reactions by choosing one of seven emojis, thereby mechanizing our feelings about our friends and simplifying them to be measurable, machine-readable, and calculable. Community and cohesion between human beings are not of value to an informatic communication system. The result, says Geoghegan, is "a subtle but profound dislocation of communications from sites of social struggle to structures of technocratic governance."[20]

In terms of voice synthesis, we see this play out quite literally. A simulation of the human voice is created informatically, and largely for the purpose of facilitating human-machine data transfer. This apes social communication, but it is *not* social communication. Furthermore, any human expression that voice synthesis simulates (it took almost 100 years of development before it could mimic any expression at all) is almost always for the purpose of getting more data into the system, not for negotiating understanding among human beings. Although voice synthesis can be used as an assistive technology for persons with disabilities of oral speech or vision, and also as an artistic medium, the most frequent experience of voice synthesis for many of us is of a voice avatar that serves as an informatic interface to our interactions with a corporate entity. As the communication philosopher John Durham Peters has eloquently summarized, "The pooling and analysis of [information] creates an implied I that is disembodied and all-seeing. . . . Information is a form of knowledge that rearranges the significance of everyday realities, sapping them of substance."[21] The informatic (synthesized) voice wrings human expression out of vocalization and encourages our everyday interactions to be with a corporate body rather than with other people.

Although it imitates the interaction of human beings communicating, the informatic voice serves a different purpose, often driven by corporate values of profit and efficiency, rather than human values of sociality and empathy. The informatic voice is not the human voice; it is the interface between human beings and the information that can be processed within networked electronic systems. Over the course of the twentieth century, actual implementations of voice synthesis often came about as a means to make the communication of information more time and cost efficient. As computers' ability to process more and more data quickly and cheaply ensured their adoption across more and more industries, the need for interfaces that allowed more people to easily interact with computed output became pressing. From automated stock prices, to automated caller assistance, to talking cars and airplanes, to the earliest smart-home appliances, voice synthesis often came into people's everyday lives not in the form of friendly or frightening talking robots, as in the movies, but as the disembodied and distant purveyor of what was often important and time-sensitive or on-the-spot information. In this way, the deployment of voice synthesis technologies for the management of everyday life caused the informatic voice itself to seem commonplace.

As this book documents, voice synthesis was domesticated long before Siri spoke from the iPhone, and users of voice synthesis technologies were already disciplined to accommodate the limitations of voice synthesis because the information spoken through synthesized voices was often important. That is less the case now. Our devices typically provide less information than we actually would get from other sources. Alexa will read "something" that it "found" on the web in response to many questions we might ask, but we don't know how it determines which "something" to read.[22]

Some research has attempted to prove that human beings are "hard wired" to react to synthesized voices as we do to "real" human voices. In contrast, this book troubles the oft-repeated dictum that voice interface is the most natural and desirable way to access information. As one engineer described it, the industry's ultimate goal is "computer-generated speech that sounds so natural, so human, that it is indistinguishable from that of a real person."[23] We might interrogate the assumptions about why an indistinguishable machine imitation of human speech is such a sought-after benchmark.[24] Lawrence Rabiner, a well-known speech synthesis researcher at AT&T, answered a variation of that question with an unsurprising appeal to the informatic values of data transfer and efficiency: "Clearly, human-machine interactions will generate enormous paybacks that justify the effort that has gone into building such machines. These paybacks will enable us to do our jobs faster, smarter, better (depending on the task being tackled), and, in some cases, with less frustration at waiting for human attendants to help answer our questions or solve our problems."[25] Instead, many examples in this book show that in fact, voice synthesis technologies have often reduced access to information, in terms of both who has access and how much and what kind of information are made available to access.

At the same time, the spread of voice interaction technologies means a reduction in our experiences of vocal diversity, an important component of social cohesion and empathy. Voice synthesis technologies have benefited people as well, as the case studies in this book will also show, but they are now increasingly implicated in our "age of surveillance capitalism," with corporations relying on voice synthesis technologies' simulation of voices to trigger our responses for the purpose of collecting our voices as data.[26] To understand some of the risks that synthesized voice applications may pose to our well-being, I reveal the significant difference that embodiment makes to the production of the human voice, as well as to the way that we negotiate our social worlds based on the sounds of each other's voices. The informatic voice, by comparison, is often an act of corporate ventriloquism that attempts to sound like a unique and dynamic human body but is instead an inflexible simulation, one often programmed to manipulate us to increase corporate profits.

VOICE AND SYNTHESIS

Voice and *synthesis* are both historically contingent categories that warrant some definition. This book focuses on the *sounds* of voices, those of people and of talking machines, as meaningful in their own right; this is why I have chosen to use the phrase "voice synthesis" rather than the more standard "speech synthesis." It is easy to conflate voice to speech and speech to language, and then focus on the meaning of language at the expense of what the voice itself reveals. As the sociologist Anne Karpf notes, "We raid speech for its semantic meaning, and then discard the voice like leftovers."[27]

Our voices are one of the primary ways we have of trying to express our internal experiences to others. Yes, language and movement play a part, but sound alone conveys so much—the difference between anger, sarcasm, and jest; feelings of fear, compassion, and vulnerability; quirks of personality;

feats of artistic beauty; extremes of rage and joy, as well as equanimity and dis-interest and despair; markers of cultural identity, gender identity, and social status; and conditions of the body, revealing health, illness, age, intoxication. When a technology deployed by an institution is given a voice interface, we are not experiencing the subjectivity of a technology that expresses emotion and experience through the tones and timbres of its voice. Rather, we are experiencing an act of ventriloquism, in which an institution—typically a wealthy and powerful corporation—"speaks" through the voice interface. For most of the development history of voice synthesis, the telltale acoustic artifacts of the electronics used to simulate the voice acted kind of like the moving lips of a lesser-skilled ventriloquist; you were aware of the simulation as a simulation because of its crudeness. Even now, synthesized voices can't simulate the full range of embodied human vocal sounds, but we accom-modate their quirks, project onto them our experiences of the voices of real people, and have expectations about both people and machines changed in the process. As computer-processing power has allowed for increasingly sophisticated data-processing techniques, synthesized voices have increased in naturalness and expressiveness, which has made it easier for us to ignore the fact of their simulation.

In this study, I use the word "voice" denotatively to mean "sound pro-duced by the vocal organs, esp[ecially] when speaking or singing, and regarded as characteristic of an individual person," or "synthesized voice" when generated by a synthesizer.[28] I also use the word "timbre," which means the character or quality of a musical sound or voice, as distinct from its pitch and intensity.[29] Timbre has been called "the psychoacoustician's multidi-mensional waste-basket category for everything that cannot be labeled pitch or loudness."[30] The human voice consists of complex acoustic wave patterns. Every sound produced through the human vocal tract includes many waves with different frequencies and amplitudes that vary rapidly, with patterns often stabilizing for only 10 milliseconds or less. One way to think of timbre is as the shape of the waves. The flexibility of the human voice comes from the mind-body's ability to vary the shape of the vocal tract. Unlike instru-ments whose shapes are rigid and relatively fixed, the human vocal tract is dynamic and changes in shape; for example, moving the tongue, lips, and jaw can change a voice's timbre. As we will see, pitch has been the primary measurement from which voice synthesis was developed. Pitch, the degree of highness or lowness of a tone, is actually our perception of the rate of vibration, called the "frequency" of a sound wave, with shorter frequencies resulting in higher pitches. Two tones of the same frequency, or pitch, can still sound different (imagine the same note played on a piano and a guitar), and this difference lies in the timbre of the voices or instruments producing the tones. When referring to sounds produced by the vocal organs of other animals, as well as certain nonspeech sounds from the human vocal apparatus, I use the term "vocalization."

Voice and *voice quality* are usually the two categories used to differentiate between the sounds produced by a vocal system and the perception of those sounds, although how those categories are defined varies from researcher to researcher. As the speech scientists Jody Kreiman and Diana Sidtis note, different disciplines focus on different aspects of voice, with engineers, for example, most interested in the acoustic waveform that correlates with vocal sound, while psychologists often focus on perception rather than the physical production of the voice.[31] If we define "voice quality" generally as "an interaction between a listener and a signal," with the listener's perception shaped not only by the auditory task and characteristics of the stimuli, but also the listener's background, it becomes clear that what one "hears" in their perceptions of voice quality is largely cultural, learned, and negotiated.[32] This is a critical phenomenon to tease apart, as our assumptions about our own voices, about the essential characteristics of others' voices, and about beliefs that a unique voice emerges from a specific body carry sway in the way that voice synthesis is deployed and in how we may respond to it.

This book's analysis spans the historical development of a group of technologies that I refer to as "voice synthesizers" because they *generate* or *concatenate* the waveforms of audio signals to simulate the human voice. I differentiate technologies that generate waveforms from those that primarily manipulate waveforms performed by a human body in real time, although the difference is somewhat subjective, even from an engineering standpoint, as signal processing is the basis of all these technologies. For mostly practical reasons, I do not cover the development of the many technologies used for voice effects in popular music, including talk boxes like Sonovox, processors including Auto-Tune, and vocoders that manipulate microphone input, such as those developed by Robert Moog and his successors.[33] These technologies are close relatives of those discussed here and, in the case of talk-boxes and the first vocoders, even antecedent. However, the focus of my study is on the through line of technologies that lead specifically to computer voice interfaces including today's digital voice assistants.

Another consequence of this scope is that in the examples of popular media that I discuss, I do not cover the full spectrum of talking androids and instead mostly limit my discussion to examples of computers and consumer devices. Androids differ from computers in that they often simulate facial expressions, eye movements, and gestures to communicate, in addition to voice. While the seemingly common dream for robotic companionship certainly contributes to the cultural imaginary for talking computers, my study doesn't attempt to trace this influence in any systematic way.[34] Furthermore, I have chosen examples largely outside of the popular science fiction films and television that are typically invoked when discussing talking machines. Stanley Kubrick's *2001: A Space Odyssey*, *Star Trek*, and *The Jetsons* do make brief appearances, but I have focused on domestic comedies and lesser-known representations rather than the science fiction often beloved of computer engineers; this is to show that people have posed challenges to the status quo of mundane computerization as often as science fiction has portrayed its power. I have identified comedy as particularly illustrative of receptions to voice synthesis across time because of its social function as a means to help audiences process contradiction and question authority, as I discuss further in chapter 2.

I use the method of the historical case study to tell the story of voice synthesis development. Although the technical development of voice synthesis has been evolutionary and iterative, it pops into the public imagination only when there is an implementation of a working product. I chose to highlight six of these products, recognizing that some developments that engineers might see as important have been left out. The media historian Lisa Gitelman explains the logic of the case study for histories of media technologies. She defines media as "socially realized structures of communication, where structures include both technological forms and their associated protocols, and where communication is a cultural practice . . . of different people . . . sharing or engaged with popular ontologies of representation."[35] She asserts that because media are both technical and cultural, media histories must embrace the complexity that entails. We cannot tell the stories of "isolated geniuses working their magic on the world"; rather, we must reveal particular technologies as particular sites for historically and culturally specific experiences of meaning.[36] She offers a cogent argument for why the case study is the best method for achieving this kind of technocultural analysis, advocating for investigations of media technologies when they emerged, as these are the points at which a media technology's "job" of representation gets constructed and questions about whether and how it does that job surface. This is when technologies are most openly revealed as "socially embedded sites for the ongoing negotiation of meaning."[37]

The job of voice synthesis, in the words of many of its developers, is to represent the human voice as intelligibly and naturally as possible. This has been the stated goal of engineers throughout the twentieth century and continues in attempts to make synthesized voices increasingly affective. But every assumption embedded in this goal demands our interrogation. As each new iteration of voice synthesis was introduced to the public, it was used and resisted, debated in the press, and engaged in processes of mutual shaping between the public, engineers, political economy, and the media.[38] Delving into new products in the evolution of voice synthesis enables me to attempt an analysis that takes seriously the relationship between engineering goals and cultural dynamics, to explore what the technological representation of the human voice was made of, materially and culturally, and to evaluate the significance of patterns and ruptures across time. In doing so, I hope to provide readers with the motivation to resist the most exploitative voice synthesis technologies. I also hope to inspire or renew our appreciation for the messy beauty and extraordinary instrument for negotiating interpersonal understanding that is our human body's own voice.

1 THE VODER (1939)

"The world has arrived at an age of cheap, complex devices of great reliability, and something is bound to come of it," an article in *Life* magazine promised readers in September 1945. From radar to Plexiglas, air conditioning to nuclear energy, *Life* featured not only people in its famous images, but also the products coming out of wartime science and engineering, reassuring its more than 13 million weekly readers that the American ingenuity that was winning the war was also ushering in a prosperous future.[1] One of these products was a talking information machine.

The article was Vannevar Bush's prophetic "As We May Think," published in July's *Atlantic Monthly* for the intelligentsia and excerpted and illustrated in September's *Life* for the masses. Bush, the most politically powerful scientist in the US and dubbed "the general of physics" by *TIME* magazine for his role in managing the development of wartime technologies including atomic weapons, was already well known to both *Atlantic* and *Life* readers. "As We May Think" was a revision of ideas that he had been chewing on as he observed the abundance of information being generated by the work of scientists throughout the war. Although information management was not new, Bush brought its problems of scale to engineering.[2] His essay argued for the development of *electronic* systems to enable improved storage, retrieval, and use of the world's increasingly unwieldy amount of information. One of these hypothetical systems, a personal information organizer that Bush called a "memex," captured the imagination of a young naval radar technician named Douglas Engelbart, who read the essay in a hut in the Philippines, where he was stationed at the end of the war. Engelbart would go on to pioneer many elements of what came to be called graphical user interfaces (GUIs) for computers.[3] But in 1945, the memex was only one of the information machines that Bush proposed. Bush speculated about possibilities for a range of information and communication devices based on technologies already in development in the 1940s, including miniaturized mass storage, rapid facsimile transmission, and voice synthesis.

For *Life* readers, these ideas were illustrated as a trio of amazing near-future machines, including one described as a "supersecretary of the coming age" that would "talk back."[4] It was illustrated to look like the mechanical cash registers of the day, with a happy face of two protruding canister eyes, a round speaker nose, and a smiling mouth of microfilm above a body of typewriter keys and a microphone trailing from a cable at its umbilicus. "It is somewhat similar to the Voder seen at New York World's Fair," stated the caption, while Bush's description in the essay was more detailed: "At a recent world fair, a machine called a Voder was shown. A girl stroked its keys and it emitted recognizable speech. No human vocal cords entered into the procedure at any point; the keys simply combined some electrically produced vibrations and passed these on to a loudspeaker."[5] The opposite process, vocoding, would also free users of needing to rely on a stenotype, a human-operated intermediary that Bush seemed bizarrely suspicious of when he described it as a "disconcerting" device requiring "a girl [who] strokes its keys languidly and looks about the room and sometimes at the speaker with a disquieting gaze." Presaging the tens of thousands of female information workers who would be displaced by new information technologies in the decades to come, this "supersecretary" device did away with the girls altogether. It could speak by itself.

Electronic voice synthesis did not begin from the goal of creating an interface for information machines, but that was one of the first practical suggestions for its use. When the engineers at the American Telegraph and Telephone Company (AT&T) first created electrical models of our vocal organs out of circuitry and current, they had been tasked with looking for ways to expand the telephone network and make it more efficient.[6] Their mandate was the corporate goal of AT&T: one policy, one system, universal service. Within that charge is the seed of the monopoly that the US government would first sanction, and later break up, but in the 1920s and 1930s, expanding domestic service and solving the physical barriers to trans-Atlantic telephony were why AT&T's researchers experimented with trying to get understandable speech sounds by manipulating electricity. Bell Telephone Laboratories was organized in 1925 to consolidate research and development for AT&T, giving its staff leeway to experiment.[7] Almost a century before Siri was integrated into the iPhone, AT&T debuted electronic voice synthesis at the 1939 World's Fair in Queens, New York, in the form of a keyboard-operated machine called the Voice Demonstrator, or Voder for short, a product of the labs' research.

Thanks to radio, magazines, and the media surrounding the fair, the Voder was famous for a time. AT&T framed it as a scientific enterprise, though it was often personified as a "he." This chapter shows how the Voder, a talking machine, serves as a hinge between the interwar *Machine Age*, with its industrial focus on electrification, assembly line production, and corporate growth, and our *Information Age*, characterized by mass digitization, pervasive use of information and communication technologies (ICTs), and the accumulation of political-economic power through the control of data and the infrastructures for its storage and processing. Although not a company generally associated with today's digital voice assistants, Bell researchers laid the groundwork for all the voice synthesis that has come since, embedding their assumptions about the mechanical nature of the human voice in all the technologies that would follow.

INVENTING THE VODER

Throughout the twentieth century, Bell Labs' speech scientists and engineers traced the lineage of their efforts to mechanize speech back to von Kempelen. This wasn't mere nostalgia; claiming antecedents like von Kempelen's was a key warrant in the argument about the body on which voice synthesis is based. In a 1972 article for *Scientific American*, James L. Flanagan, head of the acoustics research department at Bell Labs, defined speech as a "specialized acoustic code" created by "gestures" of the vocal tract. Researchers had been working to understand this code for almost 200 years, and Bell researchers, according to Flanagan, had made the most progress in doing so.[8] By defining speech as a physically produced code, voice is made ahistorical, objective, and measurable, amenable to standardization and replication. Deliberately placing the Voder within this lineage of technology highlights assumptions that engineers held about the human voice: that it could be mechanized; that the impulse to do so was a prerogative of scientific rationality (and engineering expertise); and that success would inevitably result in a social good like the telephone. Of course, providing the Voder with a long heritage also gave the company bragging rights, as exemplified in its own *Bell Telephone Quarterly*: "During the last two centuries, scientists have been endeavoring to produce speech synthetically. It remained for Bell System research specialists to be the first successfully to achieve a complete result."[9]

The first efforts at electronic voice synthesis were made just after World War I. In his brief stint at AT&T (1919–1921), the astrophysicist John Q. Stewart conducted research that he later published in the journal *Nature* as "An Electrical Analogue of the Vocal Organs," an article describing a circuit design for producing sounds like the vowels in English by substituting electrical current for the body's breath: a varying periodic pulse generator to simulate voiced sounds, and a broad-band noise generator to simulate voiceless sounds.[10] In the body, there are three ways that sound is generated by the vocal system: voiced sounds are produced by forcing air through the glottis, causing the vocal folds to vibrate; fricative sounds are produced by forcing air through a constriction in the vocal tract, creating turbulence in the air flow; and plosive sounds are produced by building up air pressure behind a closure in the tract and then abruptly releasing it. "Whether the vowel was whispered, sung, or spoken depended upon the manner of making the interruptions;" explained Stewart, "while what particular vowel was produced depended upon the adjustment of the resonant circuits."[11] His experiment could not replicate any of the explosive consonants due to "obvious difficulties of manipulation," and he granted that there was "much room for improvement with respect to the naturalness of the electrical voice," but Stewart proved that it was possible to use electricity not only to convey speech waves, but to create an imitation of them.[12] Increasingly sensitive oscilloscope technology would help show that electrical hiss and human vocal air turbulence could be mathematically equivalent, even if they sound quite different in timbre to human ears. The electric voice that Stewart was able to generate was monotonous, but observers were able to identify the sounds as imitative of human speech about half the time. Stewart identified the "really difficult problem" of artificial speech was not in constructing the device, "but in the manipulation of the apparatus to imitate the manifold variations in tone which are so important to securing naturalness."[13] As with its mechanical antecedents, from von Kempelen's to Wheatstone's, the excitation of the breath, or Stewart's analog electrical "breath," was only the beginning of recognizable speech sounds. The function of the various parts of the throat and mouth to manipulate that air into the sounds of speech was not well understood yet. In the next decade, the Voder would demonstrate Stewart's idea to the world.

The inventor of the Voder was Homer W. Dudley, who spent his entire professional career with AT&T, starting in the Western Electric engineering department in 1921, shortly before it became Bell Telephone Laboratories in 1925. Dudley was a colleague of Harry Nyquist, the Bell engineer who formalized the first important step toward quantifying information as a transmittable quantity. In 1924, Nyquist published "Certain Factors Affecting Telegraph Speed" in the *Bell System Technical Journal*; the paper showed that the number of independent pulses that could be sent through a telegraph line per unit of time is limited to twice the bandwidth of the channel, thereafter called the Nyquist Rate. Nyquist's theory was based on the two binary states of telegraph signals, open and closed, which is why it became foundational to digital information theory later. In the meantime, Nyquist and his colleagues were attempting to work out how to bring the unrulier voice waves of analog telephony in line.[14] Dudley's research into the characteristics of speech was part of these efforts.

The quest to provide transatlantic speech communication captured Dudley's imagination. In 1924, permalloy, a Bell Labs innovation, increased the bandwidth of undersea telegraph cables to 100 hertz (Hz), enabling a fourfold increase in capacity to about 400 words in Morse code per minute.[15] However, 100 Hz was not nearly enough bandwidth for transmitting speech, which required about thirty times that amount. Dudley, a telegraph man, made an analogic leap. He reasoned that the tongue was really just a telegraph key modulating a "carrier wave" emanating from the glottis, and this carrier wave should therefore be transmittable by a few tens of hertz.[16] In working through this analogy, Dudley experimented with electromechanical means of producing speech and published his first results in the *Bell Laboratories Record* in December 1936. "In ordinary telephony we move a sound wave electrically from one point to another by direct transmission," he explained, "but in the synthesizing process, only the specifications for reconstructing the sound wave are directly transmitted."[17]

Dudley filed a US patent for a "Signaling System" in the same month as the article's publication.[18] The object of the invention was to reduce the frequency range required for the transmission of speech signals, a necessary step toward Dudley's hopes for submarine telephony.[19] The patent explains in more technical detail the assumptions that Dudley held when he made the analogic leap from the vibrating air of vocal waves to the manipulated electrical signals of a synthesized voice. He began with the basic understanding that speech does not need all the frequencies included in the waves of a human voice to be intelligible as speech. This assumption came from research being conducted at Bell Labs in the new field of psychoacoustics.[20] Specifically, experiments being conducted just when Dudley started with the company had found that "normal [adult human] ears" varied little in their sensitivity to audio frequencies.[21] Research like this supported the idea that redundancies could be dropped in the process of turning voice waves into electrical signals for transmission down a telephone wire because these redundancies weren't heard anyway; it also supported Dudley's ideas that a synthesized voice need not contain all the frequencies of a human voice to create a simulation of understandable speech.

From this basis, Dudley next defined the vocal system as consisting of either fixed or variable parts. Variable parts are those that change to make different sounds, such as the lips or tongue moving. Fixed parts are those that don't change, such as the throat, and, interestingly, the vibration of the vocal folds when making voiced sounds. Dudley reasoned, "The fact that the vocal cords are always used in the voiced sounds is a fixed feature as is also the fact that they always vibrate in the same buzzerlike way as regards the presence of a fundamental frequency and all of its overtones."[22] The fact of conflating air with electricity as the carriers of speech allowed Dudley to make this abstraction. While the body is used here as a model, embodiment hardly gets in the way. Anything that might affect the fixed parts—hormonal changes, physiological quirks, or even vocal training—is bracketed in the service of engineering. There is only the quest to create "intelligible," speech-like sounds. The dynamism and diversity of human voices are irrelevant to this goal. Measurement is necessary for mechanical and computational simulation, and standardization is necessary for measurement. Of course, standardization often involves a judgment about what is normal. "The part of the vocal system corresponding to the fixed circuits of the oscillator is the same from man to man," Dudley concluded, using the perceived average of white and male bodies to stand in for every body.[23]

But what of the variable parts—the random oscillations of air/electricity that produce the unvoiced sounds? They are defined by Dudley as changes from the normal, fixed system, or *modulations*. Dudley, as had Nyquist, began to anticipate what Claude Shannon would formalize a few years later in his landmark paper "A Mathematical Theory of Communication."[24] The synthesis of speech needn't include the transmission of the entire voice wave to be understood as speech, only the *differences* from a norm needed to be transmitted to reconstruct the message at the other end of the transmission channel; these differences are the information. Just as defining standards for "average" hearing allowed engineers to figure out what could be left out of telephone signals, the idealization of an "average" body allowed for the simplification of speech signals for synthesis. In Dudley's words, "Advantage is taken of the fact that much of the information ordinarily transmitted is of an invariable or predictable character due to the general uniformity of the speech producing organs from person to person."[25]

The speech quality wasn't yet good enough to pursue incorporating Dudley's methods into the telephone system, but it did hold promise as a demonstration of what Bell scientists and engineers were capable of. Dudley worked on designing a complete synthesizer, a machine that produced intelligible speech without using any human voice or vocal recording as input. When the proof of concept was operational, management decided that this machine, now called the Voder, would join the other exhibits in the AT&T pavilion at the 1939 New York World's Fair. Dudley was granted US patent number 2,121,142 on June 21, 1938, for a "System for the Artificial Production of Vocal or Other Sounds."

The materials that made up the electromechanical Voder were very different from the air bellows lungs and rubber lips of mechanical speech machines. About 3,200 cycles of power were needed to generate the sounds. Keyboard controls, reminiscent of organ keys, and a spring-actuated foot pedal adjusted three potentiometers in the circuit of the relaxation oscillator to control pitch, with the keys setting the bottom frequency and the total octave range available for the particular voice being produced, and the pedal being used for inflection. The operator of the Voder was able to choose between an unvoiced or voiced energy source (the hiss or the buzz) by depressing a bar under the left wrist that served as a switch to "voiced." Each electrical wave from either energy source passed through a circuit that could be adjusted to create a wave spectrum that corresponded to a recognizable speech sound. This "resonant control circuit" was adjusted by pressing the ten keys of the keyboard, each corresponding to a subband of the Voder's frequency spectrum, from 0 to 7,500 cycles per second. By depressing a key, a small voltage was triggered, which increased logarithmically, according to the amount of the key's displacement, over a range of 15 decibels. As Dudley explained, the operator controlled the spectrum of the wave issuing from the Voder "in a manner analogous" to the way that the shape of one's mouth controls the spectrum of the wave issuing from "the human vocal energy sources."[26] The keys were operated "at the rates used by the tongue, lips, etc." It took a "very light touch" to operate the keys rapidly and smoothly enough to create a succession of recognizable speech sounds.[27] These waves were then amplified and radiated from a speaker.

The complicated operation of the Voder took a long time to learn. Dudley explained that Voder operators needed "a quick intelligence, the ability to control the fingers independently of one another, and a good ear."[28] Candidates drawn from the operator pool at the New York Telephone company were tested on their fingering ability, phonetic sense, and quickness of comprehension. The 320 telephone operators were winnowed down to two dozen who would train for a year to become the corps of all-women "Voder-ettes" for the World's Fair. As with musical instruments, the operators in training received one-on-one instruction and practiced for six thirty-minute sessions per day. Ten Voders were set up in acoustically treated practice rooms for this purpose. Dudley reported that it took about six months for the Voder operators to learn to form all the sounds, and another six months for them to develop the necessary technique for the speech to be intelligible.[29] The Voderettes mastered a 2,500-word vocabulary derived from Dewey's list of the most frequently used words in English, although the demonstrations were mostly scripted so the Voderettes could focus their skills.[30]

Although she was the silent partner in the Voder's demonstration, the production of understandable electromechanical speech was entirely dependent on the Voderette's movements. The Voder's speech was ultimately at the mercy of a human body for its production, even as it was meant to stand in for the human vocal system. The operator's hands had to glide at various speeds to achieve the different sounds, and her wrist had to function almost independent of her gliding fingers when a voiced sound was called for. Her feet had to control a sensitive pedal that required more delicate manipulation than just up or down. In the following passage, Dudley describes how the word "key" was produced. The passive verb constructions hide the human body's skill, which was necessary to elicit specific sounds from the machine, transferring speaking agency to the Voder itself:

> To produce this word the wrist bar, the k stop consonant key, and the proper resonance control keys are depressed simultaneously. The first operation the stop consonant key performs is the removal of the voiced energy source from the input of the resonance control filters. Further depression of the k key discharges a condenser in a resonant circuit which causes the initial damped oscillation. . . . This corresponds to the miniature explosive impulse or click initiated in the mouth of a speaker when the air-tight seal between the tongue and the hard palate is broken at the beginning of the sound k. As the k key is further depressed there is an interval of silence, then the unvoiced energy source is connected to resonance control filter [number seven] to produce a spurt of noise. Soon thereafter the stop consonant key reaches the bottom of its stroke connecting the voiced energy source to the input of the resonance control filters. By this time the resonance

control keys required to produce the ē sound of "key" are almost fully depressed and the vowel starts suddenly at almost full amplitude.[31]

The complicated movements are made clear in this description, but the body making them is hidden. The Voderette is grammatically absent from the text, leaving the machinery itself as the agent that creates the voice. The Voder is not framed as a musical instrument through which human emotion is expressed. The Voderette is not like a musician, whose interpretation of the music is a creative act. She is left out of the sentence structures that describe bits of the machine acting on their own. The woman who spent a year learning to make the machine speak is denied her own agency in the interaction, as her participation is reduced to mere technique. As described by engineers, and soon by the press and public as well, the machine spoke for itself.

THE PUBLIC MEETS THE VODER

Ahead of the World's Fair, the Voder made its debut during a presentation at the Franklin Institute in Philadelphia on January 5, 1939. The next day, variations of the Associated Press (AP) feature-length story by the Pulitzer Prize–winning science editor Howard W. Blakeslee were published across the country. "Scientists Hear Mechanical Voice," announced the *Baltimore Sun*, while Angelinos read, "New Machine Creates Speech by Pressing Keys."[32] The AP original began, "A machine which speaks, forming its own words in imitation of human tones, was shown to scientists today at the Franklin Institute." The scientific framing is key to the way that AT&T presented itself to the public. Print ad campaigns featured images of men wearing lab coats attending to the dials and wires of walls of equipment, with the caption, "The steady scientific progress of the Bell Laboratories shows in the ever-increasing quality and scope of your telephone service."[33] Inviting scientists and the press to meet the Voder at the Franklin Institute first, rather than debut alongside the spectacles of the fairgrounds, placed it within the science of the lab rather than encouraging its relationship to the fantasies of science fiction.

Nevertheless, journalists couldn't help but anthropomorphize the talking machine. And, in spite of its Franklin Institute setting, the demonstration of the Voder encouraged understanding the machine as autonomous. Even though a woman sat operating it, most articles mentioned this only in passing, if at all. One awkwardly explained, "[P]ractice is required to talk. This, on the experience of 300 telephone girls who have tried, is a period not quite so long as learning to talk with your own vocal apparatus."[34] The strange passive verb constructions allude to girls who "have tried" something, but the pronoun antecedent of the sentence is found in the preceding paragraph, which tells us that "it" (the machine) "can imitate the bleating of sheep," etc. There is enough grammatical ambiguity to further the idea that the machine is an agent—that it "learns" to talk like humans must, even though the actual investment in time is in the years' worth of practice that is necessary for "telephone girls" to manipulate the machine in such a way as to get recognizable speech out of it. As is consistent with other machinery, the humans necessary to make it perform are "operators"; and, even though the Voder looks like a small organ that would have a "player," the article describes it instead as "an oversize typewriter, with a pipe organ keyboard." This "typewriter" description assigns the Voder to the category of office and communication machinery rather than musical instrumentation, and it positions the operator to eventually be automated out of the system. As one article emphasized, the Voder spoke "without much apparent effort by the operator."[35] Indeed, in Bush's "As We My Think," his vision for a Voder-based supersecretary machine doesn't require a Voderette at all.

In this, journalists were likely following the lead of Bell engineers. In addition to the Franklin Institute demonstration, the Voder was on the radio. Recordings of this presentation provide some insight into how the machine was framed for the public.[36] It was first explained in simple machinic terms as consisting of ten electrical circuits and two energy sources that could create all the sounds of human speech. To demonstrate, the male presenter asked the operator, Helen Harper, to "have the Voder say," where "have" instead of "make" subtly foregrounds an interpretation of the machine as speaking itself, being coaxed rather than being operated. After that, the presenter seemed to converse directly with the Voder rather than with Harper, as he instructed, "Say the sentence in answer to these questions," the first of which was, "Who saw you?" The point of the demonstration was to show the

Voder as achieving vocal inflections that changed the meaning of the simple sentence, "She saw me." Using a sentence with the personal pronoun "me" furthered the illusion that the Voder itself was the speaking agent, in direct conversation with the presenter, who was addressing the machine as "you."

At times, the presenter had to acknowledge Harper as the operator, but then switched to the pronoun "he" to refer to the Voder and maintain its agency. For example, "the Voder has other voices which he can use when Miss Harper makes a simple adjustment in his mechanism." The presenter added some humor to the presentation by referring to the Voder as "still a relatively young man" whose voice may "break," providing the opportunity to show off Harper's ability to shift the pitch of the Voder's voice midsentence (although the words being spoken became much more difficult to distinguish at both the highest and lowest registers). It is interesting that the machine was gendered as male since it could simulate sounds at various pitches, though with its electrical artifacts, no pitch could be mistaken for human. Using male pronouns to refer to the Voder was symptomatic of the agency being granted to the machine rather than its operator. Vehicles operated by men—ships, airplanes, automobiles—have historically been gendered as female, and it is sometimes argued that this is because they do nothing until driven by men.[37] The Voder, with its women operators, instead inherits the gender of its male creators, or the presenter who is their surrogate, and these men's agency. The male agency associated with the control of information coming and going through technology would continue into the postwar computer age, while work denigrated as merely a *technique* needed to operate information machinery, and often done by women, would show a pattern of being low-wage and targeted for automation, as later examples in this book will show. Meanwhile, a masculine scientific authority was highlighted in the demonstration of the Voder.

The demonstration was brought back to focus on the scientific nature of the Voder within the goals of the Bell telephone system. As Homer Dudley and his colleagues stated in an article accompanying the Franklin Institute debut, the Voder was based on "the large amount of fundamental work on the physical nature of speech which has been required for the most efficient application of Bell's invention, the telephone."[38] The presentation began and ended with the Voder's scientific purpose, framing it as a serious endeavor, but it allowed a lot of rhetorical slippage when the machine was also personified with the pronoun "he" and addressed as "you."

The Voder was anthropomorphized as "he" in spite of the fact that it did not sound at all human. When asked to laugh, the Voder was made to say, "ha, ha, ha," another point of levity in the presentation, but one that also belied the limitations of the vocal simulation in terms of its fidelity to the human original. To "age" the Voder's voice, a ridiculously wobbly vibrato was added that caricatured the aging human body. The presenter, in explaining what a vibrato is, referred to it as a "peculiar quality" that made the Voder's voice more lifelike, but the vibrato was so exaggerated that it almost mocked the human equivalent.

Dudley himself likened the acoustic artifacts of the electricity that the voice was created from to any other vocal accent that might get in the way of perfect comprehension, but that's being generous.[39] Nevertheless, journalists seemed eager to write about the Voder as if it were a person. One journalist said that it "sounds like Charlie McCarthy at his worst," but most of the popular reception of the Voder in the press just passed over the fact that its voice sounded nothing like a biological human voice.[40] For example, an article in *Popular Science* stated, "He hasn't any mouth, lungs, or larynx—but he talks a blue streak."[41] And nodding to Bell's scientific framing, it added, "His creation from vacuum tubes and electrical circuits, by Bell Telephone Laboratories engineers, crowns centuries of effort to duplicate the human voice."[42] The Voder was promoted as a wonder that "whispers, shouts, speaks with the voice of a man, woman, or child, and converses intelligibly in any language, despite his slight electrical accent."[43] The press, whether out of wonder or rhetorical flourish, wanted the Voder to imitate more human vocal expression than it actually could. If you ignored that electrical accent, "the Voder can do practically anything the human voice can do . . . It may even someday attract the notice of the Metropolitan Opera," the *New York Times* speculated.[44] But even failing that, Bell Labs' president Frank Jewett provided a noble-sounding goal that brought the Voder squarely back into the mechanical realm, where expression was of little consequence: "The science of communication now has reached the point where, if everybody lost his

voice, he still might speak by punching a row of keys."[45] All these procla-
mations were hyperbole, but they fed the narrative that machinery was no
longer just for manual tasks but was now duplicating one of humankind's
unique intellectual abilities as well. What would it mean that science had
created a talking machine?

Although the press about the Voder was largely positive, there were
certainly those who questioned the wisdom of creating talking machines,
especially if people would not be able to tell the difference between the voices
of machines and those of people. Deception was the concern of a *Boston
Globe* column by Jay Franklin titled "Voder for President." Franklin face-
tiously advocated for the Voder as a "friend of the people . . . without a heart,
without a brain, but with powerful vacuum tubes and a great voice. . . . No
original idea, no taint of thought, need mar the smooth flow of his political
polysyllables. He will indeed be all things to all men."[46] It wasn't just the
fact of the Voder being a machine, however. Franklin was also disturbed by
the machine's ability to sound like anything: "Voder will speak at Atlanta
with the honey-chile twang of the piney woods, at Akron he will roll Mid-
Western r's, at San Antonio he will drawl like Billy the Kid, and in New
York City he will burst into purest German, Yiddish, Italian, Spanish, Polish, Armenian, Erse, Maori, and Urdu as the need arises. And think of the
advantage of President Voder in dealing with foreign envoys." The vision
was absurdly incongruous with the actual capabilities of the Voder, but the
fear of a populist shape-shifter is palpable in Franklin's piece. It wasn't only
the fact of intelligible and polylingual speech that disturbed Franklin, but
also the potential for imitating a range of human vocal expression, here a
marker of cultural identity that could be used to manipulate all manner of
people, with more than a hint of racism invoked in the list that only includes
non-WASPy folks as those at risk of being fooled.

Letters from the public also expressed fears of a dystopian future with
talking machines. A letter to the *New York Herald Tribune* editor published a
week after the Franklin Institute presentation looked sarcastically forward
to "the day . . . fast approaching" when devices like the Voder are standard
equipment in the home, making it "easier on the throat and on the blood

pressure to have the keys speak our nasty parts for us." The writer mused that people themselves would soon be obsolete; "Living seems to be approaching a wholly automatic state where there will be nothing further left for us to do in the world of tomorrow."[47] Notably, the letter writer viewed the Voder as a social machine, like AT&T's core service of the telephone, rather than an informational one, as Bush would do. Machine simulation of the voice, and, by extension, a machine's ability to participate in the conversations on which human sociality is largely based, was a new source of alarm (even though this was far beyond the Voder's abilities). The *Atlanta Constitution* published an anonymous editorial in a similar vein, titled "Horrors to Come!" that poked another sarcastic finger at the scientists engaged in solving problems that didn't need to be solved, citing the Voder's speech as an example: "So that within a few years man will not have to bother with waggling his tongue—he'll merely switch on a gadget and away he'll go down the conversational lane."[48]

On that same day, two separate editorials, one by "Post Impression-ist" columnist Warner Olivier, and the other anonymous, both invoked the conflicts of what was looking inevitably like another world war, the first speculating, hopefully, that "a Hitler Voder, a Mussolini and a Stalin Voder, and a stiff-backed little Fascist Voder" could be sent to "talk their heads off at each other" while the people of the world go about living in peace.[49] The second took a far dimmer view, speculating that a talking machine might be even more threatening than the world's authoritarians: "[The Voder] can promise more than Hitler ever dreamed of promising, and can promise it more persuasively. He can outrant Mussolini, he can argue down Stalin, he can outshriek anybody. This latest enfant terrible of science comes closer to being the perfect potential dictator, not merely of these United States, but of all of this much bedeviled universe."[50] As Franklin did, the writer worried about the expressive flexibility and seemingly superior rhetorical abilities that would result in fooling the mass public that advertising and media were creating. All these examples allow complete motivational agency to the machine, endowing it with power not only to simulate speech, but seemingly to construct meaningful language and vocal expression. They speak of the Voder as if it can speak entirely for itself, and they don't trust it to do so in humanity's best interests. Concerns over industrial automation were now being directed toward intellectual automation, echoing concerns that had already hit the silver screen.

Charlie Chaplin's 1936 film *Modern Times* is remembered for its satire of the Machine Age. The scene of Chaplin's Little Tramp character being sucked into the wheels and gears of a factory's machinery remains iconic nearly a century later. But the film also presents a vision of mechanization in which information relayed through electronically mediated speech wields power over the individual. In addition to the industrial automation of the body's motions to scientific management, the body's voice is also subject to industrialization. Disembodied voices of authority are mediated electronically, while the laborers are silenced. The film was made almost entirely *without* soundtrack dialogue, reminiscent of the silent film era that had been rapidly overtaken by synchronized sound technologies almost a decade earlier. This helps highlight the fact that the spoken word that *is* in the film is electronically mediated. In the famous opening scenes, the Little Tramp works on an assembly line while the company president works on a jigsaw puzzle at his desk. It seems the president's only job is to occasionally look in on the factory through an electric screen in his office and then buzz the foreman below to demand that he increase the speed of the lines. The president has omniscient views of everything going on in the building via his electronic screen, even surveilling the Little Tramp on a cigarette break in the bathroom. The president also speaks *through* the screen, as if it were today's networked video conferencing software, though even closed-circuit television had not yet been invented in 1936. The president's voice commands at a distance. His authority doesn't require any face-to-face interaction thanks to the audiovisual mediation facilitated by electronic technologies.

Even more interesting is the way that electronically mediated speech frames the presentation of an automatic feeding machine, brought to the company president for consideration by a group of men who never speak themselves. The lead man, accompanied by two lab-coated assistants, enters the office and places a case down on the president's desk, opening it to reveal a windup gramophone. Once the needle arm is placed, a voice from the recording explains: "This record comes to you through the Sales Talk

Transcription Company, Incorporated; your speaker, the mechanical sales-man." The recording then introduces the man who wound up the gramo-phone; he is actually the inventor of the feeding machine being sold, and yet he still does not speak for himself. The metareferential introduction is meant to highlight the absurdity of using "the mechanical salesman" to speak for the product, as well as the manipulations of advertising broadly. The recording parodies many of the tropes of sales speak, including promising enormous benefits and describing features in futuristic-sounding language with a hodgepodge of nonsense—"aerodynamic body," "electroporous ball bearings," "synchromesh transmission." The Tramp is chosen for a demon-stration of the feeding machine, which naturally malfunctions, causing a pie in the face and much else. It is not only the automation of the assembly line that dehumanizes the Little Tramp in this factory, but also communication technologies, which Chaplin anticipated would automate personal commu-nication itself. In this example, the automation of information transmitted by recorded voice anticipates the increasing remote control of vocal informa-tion. It's only a step away from an autonomous talking machine.

It would remain for future generations to determine if a talking machine would become a supersecretary or a means of political manipulation and surveillance (or both). The Voder was a complicated and unwieldy analog technology, an experiment billed as scientific progress, unavailable to con-sumers and serving no practical purpose; however, because it was a machine that could imitate speech, a powerful synecdoche of human beings even if the voice wasn't very human sounding, the response to it anticipated a host of questions about machines simulating human intellectual abilities, which would have to be addressed when "electronic brains" were introduced to the public a decade later. Furthermore, the analogies that engineers used to describe and explain their talking electronics, based as they were in the human body and even human psychology, primed the public to think about electronic machines in human bodily and personified terms.

Although thinking about humans in machinic metaphors and machines in human metaphors was not new, the nature of electricity gave machinic personification a new inflection. In the Industrial Revolution, pistons and arms, gears and knees, and steam and blood retained some sense of visual relationship to one another, at least as an aid to understanding how machines worked. Electricity, though, that mostly invisible life force that caused such ontological angst for the Victorians, initiated a progressive "black boxing" of technology, in which the abstractions of mathematics that made the machine work were buried deeper and deeper in the apparatus, an analogy to the human brain itself, as we will see in chapter 2. In the meantime, the 1939 New York World's Fair's promise of the wonders of "The World of Tomorrow" provided the context in which the Voder contributed to normalizing the benefits of electricity, even to the point of turning the popular imagination of electric life from Frankenstein's monster into a companionable, and talkative, robot.

THE WORLD OF TOMORROW

The stylized outline of a man's face, rimmed in gold, spanning nearly two stories inside the AT&T pavilion at the 1939 New York World's Fair, announced the "World of Tomorrow" like an ancient oracle, speaking with an electronic accent.[51] The Voder itself stood on a platform below, a nondescript half-cylinder desk when viewed from the audience's vantage, at which a female Voderette sat to give a short demonstration. Every day for eighteen months, between April 1939 and October 1940, World's Fair audiences heard the Voder speak, complete with varying pitches and inflections, as well as the vocalizations of farm animals for a bit of a laugh. The most expert of the Voderettes could even coax the machine to sing a few bars of "Auld Lang Syne" a cappella. The press often reported audiences' delight and amazement at this new talking machine, but famed essayist E. B. White, writing for the *New Yorker*, was dubious: "I remember the girl who sat so still, so clean, so tangible, producing with the tips of her fingers the synthetic speech—but the words were not the words she wanted to say, they were not the words that were in her mind."[52] White had no interest in ceding human agency to an electrical device, or, indeed, in any electrical simulation of the organic world.

In fact, White found the entire fair to be too full of electrical voices excised from organic bodies. "[I]n Tomorrow, most sounds are not the sounds themselves but a memory of sounds, or an electrification," he explained. "In the case of a cow, the moo will come to you not from the cow, but from a small aperture above your head."[53] Although he enjoyed eavesdropping on the long-distance telephone calls that were being given away by lottery in the AT&T pavilion, he generally found that "there is very little joyous song in the Fair grounds. There is a great deal of electrically transmitted joy, but very little spontaneous joy."[54] In this, White may be among the first critics of the *virtual*, finding the electronic mediation of sensory experience lacking in authenticity. He wasn't impressed by the abstraction—that electrically mediated information about the world was standing in for what might have previously been experienced closer to the source.

The fair was programmed to win over skepticism like White's. If the Chicago World's Fair of 1893 was electricity's christening, then the New York World's Fair of 1939 was electricity's debutante ball. Planned by a group of prominent New York City businessmen, the fair was a 1,200-acre promotional campaign for consumerism meant to restore the progress ideology of a nation racked by the traumatic triumvirate of a past world war, an unstable economic present, and an uncertain future, given the political volatility that was shaking Europe.[55] Planning documents show that exhibits were to fall within three general categories: goods, comfort, and welfare, the latter including education, recreation, art, and entertainment as subcategories.[56] Upon its opening, three of the fair's seven official zones highlighted the category of goods, with divisions dedicated to communication and business, production and distribution, and transportation taking up a good third of the total footprint (with amusements occupying another third). While the countries of the world did provide exhibits, most of the fair's real estate was taken up by American corporate pavilions. Steel, glass, plastic, and other industrially processed materials were everywhere in evidence of Machine Age industry as well as aesthetics, but electricity was the real star. Electricity was promoted everywhere.[57]

Electricity was a key source of industrial power, but it was also the bedrock of consumerism—extending shopping hours with electric light, increasing advertising through electronic media, and enabling cheaper mass-produced goods via new manufacturing technologies, from Borden's electric cow-milking merry-go-round to the bacon-packaging assembly line at the Swift Premium Meats pavilion. Exhibits at the fair promoted consumerism as an American ideology.

Electricity promised cleanliness, efficiency, and even freedom. One exhibit in the Westinghouse pavilion featured more-than-life-sized depictions of yesterday's "world without electricity," embodied by mannequin women holding candles for lighting, toiling over wood-fired ovens to cook, carrying ice buckets for refrigeration, and kneading laundry over a washboard for cleaning. At the end was "today's electrical freedom," embodied as a smartly dressed female mannequin holding the hands of a little white girl on one side and a matching little white boy on the other. Swinging back and forth above spectators was a pendulum reminding them that "electricity saves time" as they were ushered up to a second floor on Westinghouse Electric stairways. Spectators discovered a model "planned electric kitchen" for any size living situation, from efficiency apartment to suburban home. The high-gloss white of the cupboard faces and appliances with their chrome accents sparkled with the promise of clean efficiency. Mirrors installed in each model reflected spectators' faces back to them. They saw themselves in the middle of all this shiny electric freedom. Even outside at night, in the dark, an enormous marquee of lights blinked "BETTER LIVING," reinforcing the pervasive message of the fair.

Global communication was also part of the consumer ideology. At the AT&T pavilion, prominently located just north of the Trylon and Perisphere of the fair's theme center, one wall of an outdoor courtyard paid homage to Alexander Graham Bell, with a large bust and an inscription that read, "From whose invention of the telephone has grown a communication system that carries the human voice quickly and clearly anywhere throughout the world." However, it was electricity, again, that was literally raised to a position of prominence, as a replica of the gilded bronze statue known as the *Genius of Electricity* stood atop the pavilion, just as the twenty-four-foot-tall original topped AT&T's twenty-nine-story headquarters in Manhattan. Electricity was personified as a classical winged nude, reminiscent of Hermes, standing on top of a globe. He held in his upstretched left fist three lightning bolts, and, in his right, the frayed end of a telegraph cable that looped around his arm and across his lower body.[58] The statue was comparable to another one, not far away, at the entrance to the IBM exhibit, of a classical male nude extending his gaze and both arms up toward a fresco of air transportation, an allegory of one IBM slogan, "World Peace Through World Trade."[59] IBM was exhibiting its new radio-type machine that could transmit information instantaneously between New York and San Francisco. The company would become a factor in the development of voice synthesis in the coming decades as telecommunications and data processing converged with the development of digital computers and digital signal-processing techniques.

Certainly, one of AT&T's corporate goals for the fair was to interest a broader demographic in subscribing to telephone service, and the exhibits in its pavilion served to advance that goal. Just before the stock market crash of 1929, a little more than 40 percent of US households had telephones; however, throughout the 1930s, ownership of telephones actually decreased to a low of about 30 percent. Telephones wouldn't reach 75 percent of households until the mid-1950s.[60] Even with its monopoly power, the Bell System had significant technical and financial barriers to overcome to achieve its dream of universal telephone service. Fair visitors were still getting used to the idea of having their own voices transmitted, and the AT&T exhibits provided opportunities to hear what one's own voice sounded like through a "talking mirror" and to listen in on people's long-distance telephone calls to hear what it sounded like for voices to carry across the country.[61] These experiences were meant to demystify using the telephone, and the Voder was meant to draw attention to the scientific genius of the Bell System itself. During the fair's run, it was estimated that almost 8 million people (30 percent of the total paid admissions) visited the AT&T pavilion, and at least 5 million of them witnessed a demonstration of the Voder.[62] During the eight months that the simultaneous Golden Gate International Exhibition was open on the West Coast, another 5 million people (50 percent of the total paid admissions) visited the Bell System Exhibit there, where many of the 2 million people who heard the Voder demonstration supposedly "stood in open-mouthed wonder" at the "scientific marvel."[63] Although certainly hyperbolic, AT&T produced a candid camera picture of Voder witnesses to accompany this claim, some of whom did, indeed, have expressions of wonder. But some had expressions of confusion as well.

AT&T couldn't entirely manage the public's perceptions of the Voder. The press coverage of the Voder during the fair was more animated than after its debut at the Franklin Institute the previous January. In the US, the press coverage was more limited in its geographic scope, with much of it occurring in two local papers, the *New York Times* and the *New York Herald Tribune*; however, the international press published reports about the fair all over the world. It's notable that some accounts described the Voder as "already famous" since it had received so much national press coverage before the fair opened, including articles in *TIME* magazine, the *Christian Science Monitor*, *Popular Mechanics*, and *Scientific American*. Coverage during the fair almost always personified the Voder as "he." Although billed as being able to talk like a man or a woman, the pitch shift that created the higher "woman's" voice was only a pitch shift; it was not a distinctly different voice from the Voder's regular sound. In spite of the fact that the Voder looked like a small organ (and was operated by a woman), it was not uncommon for the papers to refer to it as a robot or an artificial man.[64] In 1939, there were competing connotations of "robot," reflecting a dichotomy of beliefs about automation being either the bane of humankind or beneficial to it.

More automation could be a threat, and some people perceived a talking machine as a move toward further automating human beings, or even replacing them. By 1939, there was already a cultural imagination for mechanical people, and it was often a dystopian one. The antifascist Czech writer Karel Čapek's 1921 play *Rossum's Universal Robots (R.U.R.)* explored what might happen if a population of synthetic people was created in a factory to be used for slave labor and then revolted against their human overseers, not unlike a fleet of Frankenstein's monsters at industrial scale. The play was very popular and had been translated from the original Czech into more than thirty languages in only a couple of years, retaining the Czech word *robota*, meaning "forced labor." The hubris of Mr. Rossum creating artificial men—*rossum* means "intellectual" in Czech—reflected Čapek's skepticism of utopian ideas about science and technology, his position after witnessing the carnage of World War I. In the US, the play was popular but sometimes controversial. In a 1923 letter to the drama editor of the *New York Times*, the Pulitzer Prize–winning poet (and one-time socialist) Carl Sandburg defended the play against charges of propaganda, arguing that it asked big questions but did not manipulate the audience to come to specific answers.[65] Čapek himself told the *New York Times Magazine* that he was deeply skeptical of American efficiency, which was only "concerned with the increase of output and not with the increase of life."[66] Čapek's play was more about the technocratic ideologies of humans than it was about robots, and his *New York Times Magazine* article didn't mention robots at all, but it was accompanied by a cartoon illustration of a street full of jolly mechanical androids that filled half the page. In the US, both industry and science fiction appropriated the robot to rehabilitate it for capitalism and a new manifest destiny of space. AT&T did not intend for the Voder to be caught up in this spectacle, but the press made it happen.

The Voder was often mentioned in conjunction with Elektro, the Moto-Man, a seven-foot-tall mechanical android built by the engineer Joseph Barnett, which gave daily demonstrations in the Westinghouse pavilion.[67] Elektro didn't synthesize speech, but a series of turntables in his chest provided a vocabulary of 700 words that could be combined into phrases through the activation of relay switches. Elektro spoke slowly, in a monotone, which gave it the cadence of what we might recognize now as a "robotic voice," although it lacked electronic artifacts—exactly the opposite of the Voder. Elektro's chin moved, but not exactly in sync with the spoken words. In addition to speaking, Elektro could perform twenty-six mechanical routines including walking (very slowly), counting on its fingers, and smoking a cigarette (its head included a small bellows).

Similar to how the Voder was demonstrated, during Elektro's performances, the illusion was created that Elektro understood and responded to voice commands that its human companion spoke into a telephone handset. In reality, each word that the man spoke was registered as an electric impulse in Elektro's chest, the pattern of which triggered one of its "tricks." Elektro was described in metaphors of human biology: his electric brain "thinks," his electric eye "sees," etc., while observers were quick to ascribe psychological states to the machine ("he has a nasty way of telling people to shut up"; "he suffers from a superiority . . . complex").[68] Elektro was known to "misbehave." The wrong number of pulses might cause Elektro to walk backward when he was expected to raise his hands, for example. This seemed to be Elektro's most enduring feature in the press. Meyer Berger, a regular fair correspondent, reported that "the aluminum wonder . . . who walks and talks like a man, forgot his company manners." Something had apparently gone wrong with some "electrical intestine."[69] During demonstrations, the presenter anthropomorphized the robot, explaining, "He's very human. If you don't talk right to him he just doesn't do what you say."[70] The Voder demonstrator's ad-lib about that machine being a "young man" encouraged the same kind of sympathetic projection of human experience onto the machine. Of course, not everyone was amused by the antics of the giant android. Some responses echoed concerns expressed about the Voder, and some thought that these experiments were trivial, or even irresponsible, in light of the conflicts that were very soon to result in worldwide war. For the most part, though, the anodyne robot charmed its audience.

Westinghouse had been building android robots for a decade by the time Elektro became a celebrity at the 1939 World's Fair. In his cultural history of American robots, Dustin A. Abnet shows how Westinghouse engineers made android spectacles out of the old technology of vacuum tubes as part of a public relations campaign to promote the company's products and recruit engineers.[71] Instead of focusing on people being treated like machines or machines taking the place of people, Westinghouse promoted the idea that anyone could use an electric robot (or, more likely a modern electrical device like a Westinghouse kitchen appliance) to accomplish their individual daily drudgeries.[72] This was a domestic vision, sold to middle-class white families and meant to placate skepticism about the Machine Age like Čapek's. In advertisements, Westinghouse pointed out that unlike Frankenstein's monster, Elektro was "all kindness and geniality" and could even be operated by a "slip of a girl."[73] In the Westinghouse propaganda film *The Middleton Family at the New York World's Fair*, the school-aged son, Bud, is inspired by the Junior Hall of Science to become an electrical engineer someday rather than be cynical about joining President Franklin D. Roosevelt's Depression-era work programs (not-so-subtly tied to communism in the film); and the older daughter, Babs, dumps her immigrant–art teacher–socialist boyfriend for the hometown boy (and Westinghouse apprentice) Jim. With healthy doses of racism and sexism thrown in, the Middleton offspring learn to believe in the promises of engineering and efficiency as products of rugged individualism and capitalism, the ideological tenets of the corporate World of Tomorrow. Talking machines were the mascots of this ideology at the fair.

THE PEDRO EFFECT

The vice president of Bell Labs, O. E. Buckley, described the Voder as "a stunt outgrowth of a very serious line of research," although it was the stunt that captured the public imagination.[74] When the AP reported about the Voder after its Franklin Institute debut, the article mentioned that the machine had been given the nickname "Pedro," based on an apocryphal anecdote from Bell Telephone history. When Alexander Graham Bell first demonstrated his telephone at the 1876 Centennial International Exhibition in Philadelphia, one of the exhibit judges, Dom Pedro, the emperor of Brazil, is said to have declared in shock, upon hearing Bell's voice come out of the receiver, "My God! It speaks!"[75] Sixty years later, the Voder was a machine that was widely described as speaking on its own, in spite of the fact that a woman sat, in full view, operating it.

We might refer to receptions of the Voder as the "Pedro Effect," after the well-known human psychological response to natural-language processing (NLP) by computers that is known as the "Eliza Effect." The Eliza Effect refers to the illusion that an interactive computer system is more intelligent, complex, and capable than it actually is. It's named after the Eliza chatbot developed in 1966 by Joseph Weizenbaum, a computer scientist turned artificial intelligence (AI) Cassandra. More specifically, the cognitive scientist Douglas Hofstadter has defined the Eliza Effect as "the susceptibility of people to read far more understanding than is warranted into strings of symbols—especially words—strung together by computers."[76] The Pedro Effect precedes the Eliza Effect, but it suggests the same response to voice communication. We might define it as the susceptibility of people to attribute far more understanding and autonomy to machines than is warranted when the sounds made by machines can be interpreted as recognizable words. We'll see the Pedro Effect repeated with voice synthesis technologies to come.

In 1939, the Voder didn't speak on its own, of course; an immense amount of human skill was required to coax understandable speech sounds from the machine. E. B. White wasn't entirely correct; although the Voder presentations were scripted, the Voderettes were the only people who could have "spoken their minds" via the machine. Even more to the point, the analog nature of the machine meant that the operator could manipulate it to create some sounds of her own. It would be the only means of voice synthesis to offer something like a human being's musical expression until the twenty-first century.[77] The voice synthesis technologies to come, based on the abstract model of the human vocal system that Dudley initiated, would not require human fingers to play a voice into being, but they would also struggle to include even a modicum of inflection for several decades. Nevertheless, they would continue to habituate the public to the idea of vocal machines, even Bush's supersecretary, especially as machines developed during the impending war became key sources of stored and processed information.

2 ELECTRONIC VOCAL TRACT (1950)

It would be almost forty years before another voice synthesizer gained as wide a public profile in the US as the Voder. By then the public expected machines to have voices. This expectation was promoted to the public in the intervening years, even as applications of voice synthesis largely remained inside research and development labs. The public started to believe that machines might soon talk not only as the result of popular science fiction, as is commonly acknowledged, but also because of the way that the electronic machinery emerging in the wake of World War II, especially the stored program electronic computer, was framed in advertising, the news media, and by many industry boosters. Widely described by analogy to the human body, the new electronic *brains*, like all brains, needed to communicate. Metaphorically, computers were born as subjects rather than objects, often discussed in print as if they were people, and characterized as oracles, machines that had the power of prediction by virtue of their ability to quickly calculate probabilities. It seemed reasonable for the public to imagine that electronic brains must have electronic voices.

In fact, electronic vocal tracts (EVTs) were being built, as the human vocal apparatus continued to be measured and modeled by Bell Labs and other telephone system researchers in the hopes of increasing the scale and efficiency of telecommunications systems. Although their exigence was telephony, EVTs would provide the theoretical foundations for getting computers to talk. Voice synthesis wouldn't emerge as a computer interface until the 1960s, but the public's imagination for this was being primed by the new medium of television—not only by science fiction, but also by the science news that was a genre staple for broadcasters in need of programming, and by advertising and computer product placements in popular shows. Even though million-dollar mainframe computer installations were few in the early 1950s, television audiences were learning about them, and, by the end of the decade, vicariously debating their social power through television comedy.

This chapter picks up with speech research at Bell Labs after World War II, showing how new research instrumentation, along with increasing demand for global telecommunications, furthered the measurement and eventual digitization of the human voice. Simultaneously, the giant electronic calculators developed during the war made their public debut, including the UNIVAC computer, which was often promoted on television and shows how commercial computing was framed for the public. The audiovisual nature of television resulted in computers often being treated as characters, even in factual programming, reinforcing the idea that computers, like the Voder before, were autonomous agents with which to communicate. Television programming also became a venue for negotiating the social limits of computerization, especially within comedic genres that mocked the electronic brains as a way of questioning computation's role in society. This chapter shows that by the end of the 1950s, the idea that electronic computers were social agents was firmly entrenched through popular media. The assumption that information from machines would be shared through electronic voices was often taken for granted in these representations.

INVENTING ELECTRONIC THROATS

Speech-processing research at Bell Labs picks up where Dudley's Voder left off with the development of the spectrograph machine in 1944 for visualizing acoustic waves, an early example of signal-processing hardware.[1] Dudley explained it as displaying "the telegraph-like character of speech for the eye to view without any carrier synthesis of speech for the ear to hear."[2] Although telegraphy seems a step backward from the universal service *telephony* that AT&T focused on after World War II, the terminology conveniently enrolls speech coding in the lineage of the Bell System. Spectrograms made speech waves visible in the same way that Alexander Graham Bell's father had sought to make oral phonetics (the facial movements of a speaker's mouth) visible so that deaf people could more easily learn to speak.[3] In 1947, *National Geographic* magazine reported that deaf people were learning to "read" the sound patterns, and, "better still deaf people can use [visible speech] to improve their speech . . . by watching the patterns of their voices on the screen and comparing them with normal speech."[4]

For AT&T, though, the real benefit of mapping acoustic waves with the spectrograph was that it allowed researchers to identify more details about the channel of voice communication (i.e., the sound waves themselves moving through the vocal tract) for the purpose of figuring out what parts of the signal were critical and what could be compressed. Homer Dudley maintained his analogy of the vocal system to the telegraph, explaining, "Speech—in its first vibratory expression as muscular wave motion—is a set of telegraph signals."[5] Spectrograms allowed Bell researchers to strip acoustic waves down to their fundamentals, analyze their formants and frequencies, and reveal the "telegraphic" nature of speech, a metaphor that would later prove particularly well suited to digital signal processing (DSP).

Where the Voder's electromechanical apparatus manipulated the buzz and hiss of electricity analogous to air in the vocal system, the focus on viewing and measuring the waves in the voice signal made it possible to try to model not only what happened at the lips, but what happened in the throat, nose, and mouth. In the interim before computers were capable of modeling the vocal tract digitally, basic research focused on simulating the entire vocal tract, from folds to tongue, with electronics. In 1950, Bell researcher H. K. Dunn proposed a way to calculate vowel resonances through a cylinder and converted that numerical model into the first iteration of an EVT. X-ray technology developed at the turn of the century had made investigating the inner workings of the human throat possible, and Dunn based his calculations on published x-ray images of "subject 326," a male participant in a study conducted by Ohio State University speech scientist G. Oscar Russell.[6] Dunn said of the images: "When one examines [them], it is obvious that a mathematical treatment that would take into account all the small variations in shape would be an extremely complex and difficult matter. A simplification of some kind is necessary; yet the treatment of the vocal tract as a double Helmholtz resonator, using only the volumes of the cavities and the conductivities of the passages between them and to the open air is too great."[7]

Instead, Dunn chose the metaphor of the transmission line for his model of the tubelike human throat. He explained, "A uniform cylindrical section . . . having a plane wave passing through it, is analogous to a section of transmission line. That is, the acoustical resistance, mass, and compliance are distributed along the cylinder in the same way that resistance, inductance, and capacitance are distributed along the line."[8] He therefore modeled the human body's vocal tract as a series of cylindrical sections placed end to end like telegraph wires and used the calculations generated from this model to construct a circuit, or "electrical analog," with a vibratory energy source standing in for the vocal folds as the Voder had. It was a mathematical abstraction of human biology based on the measurements of one body deemed to be normal and then simplified to overcome the complexity of a real body's vocal system.

Dunn predicted that his first EVT could be used as a phonetic standard: "a vowel could be specified in terms of electrical constants . . . and reproduced at any time or place without the variability associated with the human voice."[9] In information theory, novelty is expensive. This was worked out by Claude Shannon, an MIT graduate from Michigan who rode out the war as a cryptologist at Bell Labs, and who published one of the most important documents in the development of electronic information theory, "A Mathematical Theory of Communication," in the *Bell Systems Technical Journal* in 1948. As a cryptologist, Shannon was immersed in ideas about the statistical properties of language, not semantics, expression, or sociality.[10] In "A Mathematical Theory," Shannon defined (among other things) how much information could be sent through a given communications channel as binary digits, or bits, per second, with "information" defined as statistical patterns in a message, not the message's meaning.[11] In 1950, Dunn was reasoning that transmitting measurements about standard, synthesized vowels instead of the sound waves produced by a body could be one way to increase the efficiency of telecommunications. The idea was far from practical to implement, but Dunn's research was undertaken in an environment of rapid innovation, where it was easy to imagine that such a solution might soon be possible.

One of the most pressing goals for both AT&T and some of the European telephone networks was still to set a transatlantic telephone cable, the problem that Dudley had also worked on. Radio transmission was the standard for global communications coming out of World War II, but telephony promised more security and stability.[12] Signal compression was needed to transmit voices under the ocean from one coast to another, and EVTs were an experimental means of developing it. The UK government was funding research similar to Dunn's, including a project at the University of Edinburgh to build and experiment with an EVT called PAT, which stood for "Parametric Artificial Talker." As at Bell Labs, the Edinburgh researchers were looking to define the critical properties of speech so that noncritical aspects could be compressed or eliminated using the statistical logic of information theory. The parameters that researchers were investigating were loudness, pitch, fricative noise, and vowel overtones, not only some of the most prominent features of a voice signal but also the most amenable to quantification, given the measuring instruments available. The developers of PAT were trying to determine if these parameters were enough to create a waveform that sounded like recognizable speech. Researchers collected data from human test subjects using analog metering instruments and then fed the resulting waveforms into the PAT equipment as spectrograms drawn with conductive ink. The various research procedures were invasively physical, with the instruments of research plunging into the subjects' bodies to measure what was happening in their throats. Reminiscent of the time and motion studies of human workers that were a hallmark of late nineteenth-century industrial scientific management, the PAT research also sought to identify boundaries between human bodies and machines that could result in system efficiencies. The goal was to engineer variation out of the system.

A 1958 television episode produced by the BBC called "The Six Parameters of P.A.T." introduced the EVT as "what can only be described as a talking machine," and demonstrated what research with PAT consisted of.[13] The machine spoke before any of the scientists, repeating the title of the documentary series, *Eye on Research*, as scientists adjusted the EVT's knobs and wires. Science and engineering were often covered by print journalism, but the advent of television provided new opportunities for audiovisual engagement, and this BBC documentary made the most of the relationship between the static PAT machinery and the research subjects' bodies, which were more interesting to watch. The documentary began with the telephony engineer Walter Lawrence explaining, "We could make much better use of these [undersea telephone] cables if we abandoned the traditional method of sending a signal that is a faithful copy of the intricate soundwave patterns [of the human voice] and sent instead signals that describe simple properties of the wave like loudness, and pitch, and overtone structure," the properties to be measured.[14] A researcher with a tube sticking up his nose explained that loudness was a consequence of the force with which air was expelled from the lungs, and that he had swallowed the tube into his esophagus to measure this force. The exterior end of the tube was attached to a measuring apparatus. He pushed a plunger to force a measured amount of air through the tube and into a small balloon at the internal end (resting inside his larynx). As he then talked with the balloon in his larynx, the amplitude of his voice wave could be seen on the attached oscilloscope. Next, he explained how researchers were investigating air being forced through the vocal folds by using a laryngoscope hooked up to a motion picture camera that captured the vibration of the folds on film, which could then be studied in slow motion playback. At Bell Labs, researchers developed high-speed cameras for just this kind of research, achieving speeds of 10,000 frames per second in the 1940s.[15]

The documentary next explained the use of a palatograph for measuring tongue position, showing a lab assistant having her tongue sprayed with a black powder, making the consonant sound "see," and then inserting the mirror of the palatograph into her mouth to show where the powder had been removed from the tongue during her mouth's motion when forming the sound. The researcher finally explained the relationship between the vibration of air in the throat and the overtones characteristic of English vowel sounds by flicking the back of his middle finger against his throat as he said "ah." The result of these demonstrations was to blur the boundary between body and machine. Measurements taken of human bodies went into PAT, and out came a rough simulation of the voice.

PAT filled six floor-to-ceiling equipment racks. It was described as having a "larynx" that was a series of knobs that could vary the frequency of waveforms, resulting in changes in pitch and having a range of several octaves. In fact, each of the parameters could be manually manipulated, but PAT also converted visual input into synthesized voice output. That is, each of the measured parameters could be input to PAT as waveform drawings that it used as instructions for producing synthesized voices. PAT's voice was about the same as the Voder's in terms of tone, but it was less dynamic and had distinct electrical artifacts. It couldn't imitate a Scottish brogue, nor could it manage implosive sounds, but neither did it require a Voderette. PAT would never prove practical for use in the telephone system, and the UK-funded transatlantic cable (TAT1) came online in 1956 with no added efficiency from EVT experiments. The research did contribute, though, to the mathematical model of the human voice that engineering was constructing, piece by piece.

Researchers at the Massachusetts Institute of Technology (MIT), the nonprofit Haskins Laboratories, and the Royal Institute of Technology in Sweden also built EVTs, improving on Dunn's model. They were primarily used in perception studies to figure out what test subjects were hearing when the EVT produced different sounds, and they were key to early research in the fields of psychoacoustics and electroacoustics. A blurb in the November 1961 newsletter of the Society for Science and the Public described MIT's EVT, noting that MIT researchers were predicting that speech synthesizers would play an "increasingly important role" as communications "between man and machine advance."[16] It was a vague prediction, but evidence that researchers were imagining applications for voice synthesis beyond making telecommunications more efficient. These three labs, in particular, were working on the long-term goal of text-to-speech output for a reading machine for the blind, on the suggestion of Homer Dudley during a meeting of the Blind Veterans Association in 1954, further discussed in chapter 5.[17]

Interested members of the public might have learned about EVTs in the popular science literature or on television in the early 1950s. They would also have become aware of the new "electronic brains," pitched as "automatic machines for handling information by electronic or electrical relay circuits" that promised to "open up vast areas of the fields of science to research," and also to "revolutionize the handling of information in business."[18] The circuits of electronic brains and the circuits of EVTs would probably not have seemed vastly different to the public. The fact that both were introduced from the earliest stages of development by metaphorical reference to the human body further connected them.

In December 1947, the *New York Times* first reported on a new "mathematical brain" that was set to be completed in the next year, the Electronic Discrete Variable Automatic Computer (EDVAC).[19] The article covered a meeting of 300 engineers, mathematicians, and scientists gathered at the Aberdeen Proving Ground, where the EDVAC was to be installed, in order to discuss its progress.[20] A booster of the project, Edmund C. Berkeley, an actuary for the Prudential Life Insurance Company, served as a science translator for the reading public when he was quoted at the end of the article explaining the relevance of this new kind of "brain": "The machine could take in the data in regard to [an insurance] policy being surrendered, look up the cash value in [the] proper table, interpolate for the premium paid to date, multiply by the amount of the insurance, deduct any loans, compute the interest on each loan, and subtract that, credit the value of any dividend accruals and any premiums paid in advance and type out the check to the policyholder in payment of the net value of the policy."[21] Getting a timely and accurate check from one's insurance company made sense to *Times* readers in 1947, even if the scientific and military applications of computing were less apparent to the public.

Berkeley published *Giant Brains, or Machines That Think* in 1949, a popular explanation of the power in these new machines, even though only six of them were operational in 1949. The book contained some algebra and schematics but was meant for a popular audience. In the preface, Berkeley explained his use of the extended analogy to human being: "I have talked of mechanical brains as if they were living. For example, instead of 'capacity to store information' I have spoken of 'memory.' Of course, the machines are not living; but they do have individuality, responsiveness, and other traits of living beings . . . Besides, to treat things as persons is a help in making any subject vivid and understandable."[22] Even as Berkeley explained his metaphor, he couldn't help but use it. That a machine with "individuality, responsiveness, and other traits of living beings" would express itself through voice seems like a given.

Berkeley was not solely responsible for the metaphor, though. It had the ethos of science and philosophy behind it as well. *Giant Brains* followed Norbert Wiener's surprise bestseller *Cybernetics: Or Control and Communication in the Animal and the Machine* (1948), which sold 15,000 copies in its first eighteen months of publication. Both books received wide coverage in the press and overlapped in their assertions about the computational nature of human thinking and the future in a world with machinery that surpassed people in certain cognitive abilities, even as they differed in optimism about that future. In the *Saturday Review*, the provocative title "Devaluing the Human Brain" led a review of *Cybernetics*, calling it a "must" read. The reviewer noted that "the artificial brains described by Professor Wiener are startling machines. They accept information and instructions and act upon them. They can even gather information as bases for decisions. They perform more reliably, accurately, and quickly than human brains."[23]

Wiener had been a prodigy, graduating from Tufts College with a degree in mathematics at the age of fourteen, and from Harvard with a PhD at nineteen. His varied interests are reflected in the fact that he taught philosophy, worked as an electrical engineer, and was also briefly a journalist before working on ballistics in World War I. He spent the rest of his career as a professor of mathematics at MIT, working on antiaircraft defense during World War II, which led him, independent of Claude Shannon, to statistical ideas that became central to information theory. "Cybernetics" was the name that Wiener gave to his new interdisciplinary study of feedback processes, seeing them paralleled in both electronic circuitry and human cognition. Wiener himself was part metaphysician, musing that men create machines in their own image, and computer brains were no exception. As the historian Ronald Kline has documented, cybernetics was a "craze" emblematic of the 1950s, cited not just by technologists but also by cultural critics as seemingly far afield as the novelist James Baldwin.[24] Much of Wiener's mathematical sophistication was lost as cybernetics was popularized in the press, but his warnings about the automation threats of what he called a second Industrial Revolution, what others since have called the "information economy," were often taken very seriously.[25]

A feature article about computers in the *Saturday Evening Post* in February 1950 was titled, bluntly, "You're Not Very Smart After All."[26] It began, "Now the scientists have come up with 'mechanical brains'—electronic monsters that solve in seconds a problem that would take you hours. They're human enough to play gin rummy, even have nervous breakdowns." The journalist interpreted scientific efforts to test the computability of specific activities as the desire of the machine itself for leisure (it can play gin rummy), and psychologized the machine's glitches (it can have nervous breakdowns).[27] He further takes cues from Berkeley in describing the computer's "five organs": its input system were the "eyes," computing unit the "brain," storage cell the "memory," central control the "nervous system," and the output system the "voice." This kind of anthropomorphic framing, oft repeated, would become tropes about computers in the media. Even as it frustrated some scientists, like Harvard's Howard Aiken, who declared that "[computers] can't think any more than a stone," the "brain" metaphor resulted in descriptions of computers as fully autonomous thinking, feeling, and talking machines, although it remained to be seen whether they would be friend or foe.[28] These fears were often discussed in the press, and they were also negotiated through the new medium of television.

STAY TUNED! FOR THE UNIVAC

The physicist John Mauchly and the electrical engineer J. Presper Eckert met at the Moore School of Electrical Engineering at the University of Pennsylvania, a center for wartime computing in the US, where Mauchly first proposed building a general-purpose electronic computer in 1942. In 1943, the US Army contracted with the school to build that computer, the ENIAC, and another, the EDVAC, in 1944. The duo left the university in 1947 to form the Eckert-Mauchly Computer Corporation, the first commercial computer company in the US, and they secured a contract to build a machine for the US Census Bureau through the National Bureau of Standards. The market for the multimillion-dollar machines was small though, and financial stability for the fledgling company was elusive. Mauchly's sales pitch was turned down over and over by the largest insurance companies and military contractors in the country, those likeliest to have needs that justified the enormous expense. Barely off the ground, the company became a division of the business machine company Remington Rand in 1950, out of financial necessity, before installing even a single machine. The computer that they were trying to sell was named UNIVAC, for UNIVersal Automatic Computer, a carefully chosen moniker that they hoped would highlight its usefulness for scientists, engineers, *and* businesses. The first was finally completed for the Census Bureau in March 1951, and a second was installed at the Pentagon in June 1952.[29]

Remington Rand was now in the business of selling computers and, like many other US businesses in the early 1950s, it turned to the popular postwar medium of broadcast television to promote its products. Television debuted at the 1939 World's Fair (along with the Voder), but, due to the war, its development and the growth of a broadcasting infrastructure were delayed. At the end of World War II, there were only a few thousand television sets in American homes; by 1951, there were 13 million—still only 10 percent of households, but this percentage was rapidly increasing. By 1955, 63 percent of households would claim to have a television set, and 85 percent would have one by the end of the decade.[30] By 1956, television ad sales reached $100 million per quarter.[31] Television was quickly ascending to become the main information technology in the US.

Remington Rand was aggressive about getting UNIVAC on television. They produced ad spots that played during breaks on the popular Sunday night panel show *What's My Line?* for which Remington Rand was a named sponsor in 1956. The dual benefits of prediction and calculating speed were stressed. The common thirty-second spot highlighted the computer's efficiency benefits, like its ability to calculate payroll and produce checks "in a flash." A more in-depth three-minute ad explained the useful application of UNIVAC's data-processing abilities to predict the weather, something that UNIVAC's inventors were actively pursuing, but also something outside human control that a broad audience could connect with.[32] Over fearmongering imagery of hurricane force winds and dangerous surf, the narrator

began by intoning, "This is weather. One of Nature's ever-changing mysteries." As the imagery changed to a computer center, the narrator introduced
"the hero of our story . . . the giant electronic brain developed by Remington
Rand: UNIVAC!" which promised to give people knowledge to protect them
from dangerous weather. Perhaps straining to make computation exciting,
the narrator used the visual of paper maps to explain how UNIVAC could
analyze past storms and then "make predictions," a phrase that he emphasized.[33] The additional key to UNIVAC's usefulness was the speed with
which it could provide these predictions, especially compared to the calculation capabilities of people. Personifying UNIVAC was another strategy for
helping it play to the camera. Panning to a young woman standing next to
the UNIVAC's printout equipment, the narrator said, "Here is where we
get the answers. You might call it the voice of UNIVAC." The silent young
woman is there to retrieve this printout, although her role is not mentioned
in the narration; UNIVAC is described in autonomous terms and given the
agency that comes with having a metaphorical voice of its own.

In addition to advertising, television was the newest medium for
political information. Remington Rand saw an opportunity here as well, as
computers' potential to calculate the probabilities of election outcomes also
generated interest. In 1947, Harry Truman became the first US president
to give a televised address, and in 1948 he became the first to broadcast an
election campaign ad.[34] The 1952 presidential election was the first to have significant television coverage. The Republican candidate, Dwight Eisenhower,
ran catchy, animated "I Like Ike" advertisements, and the Democrat, Adlai
Stevenson, broadcast thirty-minute voter information sessions. In the early
1950s, the three primary broadcast networks first established for radio—ABC,
NBC, and CBS—were competing for Americans' television viewing attention, and all three covered the 1952 conventions of both parties, programming on which each of them lost revenue.[35] Each network was looking for a
hook that would entice viewers to their premiere election night coverage. As
the industry magazine *Broadcasting* explained, "TV networks, engaged in
their first coast-to-coast reporting of a national election, employed a broad
variety of gadgets, both electronic and mechanical, as visual aids and also . . .
electronic 'brains' enlisted . . . as forecasters of final results."[36]

CBS partnered with Remington Rand to use the UNIVAC. On election night, anchorman Walter Cronkite introduced "that miracle of the modern age, the electronic brain UNIVAC." The actual computation was being done at Remington Rand headquarters in Philadelphia, but a version of the most visually interesting main console, with its blinking rows of lights, was installed against a wall in CBS's election headquarters, where dozens of men and women buzzed around at desks filled with telephones and paper and large teletype machines. Newscaster Charles Collingwood was assigned to cover the UNIVAC. He reassured the audience, "This is not a joke or a trick. It's an experiment. We think it's going to work." It did, but not without a few bumps along the way.

Collingwood interviewed UNIVAC as if "he" were a person, telling the audience that "he lives down in Philadelphia; he's one of a family of electronic brains." The printer output peripheral was described as "the way UNIVAC talks," although the communication was not without glitches. When Collingwood asked, "Can you say something, UNIVAC? Can you say something to the television audience?" the printer stayed still. Collingwood ad-libbed with, "You're a very impolite machine, I must say."[37] Unbeknownst to Collingwood, UNIVAC did have "something to say," as the computer had in fact already calculated a projected win for Eisenhower by 8:30 p.m. EST based on the data model that programmers had used. However, this was not the outcome that human pundits and pollsters had predicted, so Remington Rand's team decided to adjust their model on the fly, and CBS producers decided not to announce the prediction. By 10 p.m., though, with more actual returns verified, it seemed clear to the humans as well that Eisenhower would be elected, and Eisenhower himself had taken the stage at the Commodore Hotel in Manhattan, sounding confident that he would win. Meanwhile, Collingwood, stuck with the supposedly noncommunicative UNIVAC, told the viewing audience, "We're having a little bit of trouble with UNIVAC. It seems that he's rebelling against the human element."

Viewers were finally told around 10:30 p.m. that UNIVAC predicted an Eisenhower win, and about an hour later, Edward R. Murrow officially announced the election for Eisenhower. At 12:30 a.m., CBS cut to Philadelphia, where Arthur Draper of Remington Rand was standing by to explain "what happened to UNIVAC." Draper stated, "We had a lot of troubles tonight. Strangely enough, they were all human, and not the machine." It turned out that UNIVAC's initial calculation had been closer to the actual outcome of the election, predicting forty-three of the forty-eight US states going for Eisenhower. "We just plain didn't believe it," said Draper. "We should have had nerve enough to believe the machine in the first place. It was right. We were wrong." Even though it was a human-designed data model that actually "predicted" the Eisenhower win based on early return data, the entire evening was framed as the UNIVAC having its own secret knowledge about human events that people couldn't themselves believe. The more technically accurate details about what actually took place are hidden in the make-believe of an autonomous computer brain attempting to communicate its superior knowledge and being silenced. Even Remington Rand played along with the idea that UNIVAC was something like an oracle—a fiction that was easier to make "work" on television than an explanation that required some basic statistical literacy to understand.[38]

Some newspapers made fun of the "mishap" the next day, but reviews of the networks' broadcasts tended to focus on the "human element," generally unimpressed with the UNIVAC. In the *New York Times*, Jack Gould praised CBS for its superior visuals—forty-eight large, individual panels recording the tallies from each state that were easier to see than ABC's "old-fashioned blackboard." NBC had also used an electronic brain, a Monroe Systems Monrobot that predicted an Eisenhower win at about 10:30 p.m., but Gould found both "gadgets were more of a nuisance," although he appreciated the way CBS's anchors gave UNIVAC "a rough ride" for refusing "to work with anything like the efficiency of the human being."[39] The critic writing for *Variety* was even more pointed. A month before the election, *Variety* had reported bombastically on its front page that "television is zooming into the realm of science fiction to predict election returns . . . via . . . the use of all-electronic 'brains' which will do virtually everything but make the acceptance speech for the winning candidate."[40] In a postelection review, though, titled "Machine vs. Man," they stated that "if anything, TV's unprecedented coverage of last Tuesday's election returns demonstrated that the machine will never take the place of the human," finding that the UNIVAC and Monrobot were "expensive and awesome" gimmicks that were, nevertheless, "cold." Using explicitly emotional language, *Variety* argued that it was not machines that made for compelling television, but rather the sympathetic human anchors who were able to understand what the voter "felt in his heart" and "transmit to the millions of viewers the warmth and the spirit of America at the polls, the frustrations and despair of the vanquished and the jubilations of the victor."[41] *Billboard* praised CBS's visuals and the "calmly authoritative" Cronkite, with no mention of UNIVAC or Monrobot at all.[42] In spite of the anthropomorphizing of UNIVAC, it was no competition for human drama, according to most critics.

Although histories of UNIVAC have often pointed to its use in the 1952 election as a triumphant public debut, these and other responses show a more nuanced reception. The idea of drawing normative conclusions by collecting statistics had gained momentum in the nineteenth century, driven by exploding urban populations and developments in mechanical calculation. Social engineering facilitated by the use of population data was well established before the UNIVAC came along, but it remained controversial, of course, both socially and individually.[43] What would people need to conclude about the human condition if all human behaviors were predictable? A *Wall Street Journal* editorial worried about the future of democracy itself. "Its designers assure us that the Univac is not psychic," explained the editors, "but if it succeeds it will answer a deep psychic urge to know now what will happen later." This was not a bright future. The editors used sarcasm to make this point: "It will surely succeed if people will just eliminate the human perversity of changing their voting patterns betwixt elections. And what vistas are opened up when voters will just vote with robot predictability. With a prophetic Univac we won't have to bother to count all the votes. Indeed, if we can just improve the breed we won't even have to hold an election at all."[44]

The editorial echoed concerns about "mechanical men" voiced before the war, but it was now also asking questions about the role of statistics in society, not just electronic control. Written less than a decade after the US engaged in a war against authoritarianism, the editorial didn't ask what computing can do, but, rather, what it *should* do. In a democracy, variability is a feature of human beings, not a bug. The editorial took the discipline of data determinism to its ultimate conclusion and used sarcasm to make the point that we should reject it.

Even if less direct about saying so, election coverage reviews that focused on human elements were negotiating the boundaries between humans and machines, describing the social benefits of emotion and human variability at the same time as UNIVAC's boosters mocked the misguided "human element." Of course, not all computer proponents appreciated this strategy. A letter to the editor of the *Washington Post* criticized that paper's editorial, which had personified the machine as "Professor Univac" in defending its superiority against the "stupid humans" who wouldn't "believe him." The complainant wrote: "Your editorial . . . unhappily tends to further the mistaken notion that Univac worked out a system of election predictions. It did not. What it did was to process at high speed the election returns according to a prediction scheme worked out by human beings."[45] This simple and critical fact, one deserving of a public accounting, is lost again and again in coverage of the event and in the personification of the computer as a brain with a voice.[46]

Anthropomorphizing the computer was a useful strategy for explaining its complexity, but also for hiding it. Some people did take UNIVAC's "personhood" almost at face value. In 1955, Collingwood himself remembered the UNIVAC, "him," fondly, saying, "He fascinated people and they began to give him a personality. People called up and wanted to argue with him." Noting that many critics and commentators had taken comfort in the hiccups CBS experienced, Collingwood chided, "Well, let them sneer. UNIVAC doesn't care. He's got a steady job with Metropolitan Life and the rest of his family is well taken care of by U.S. Steel, the Atomic Energy Commission and other outfits who appreciate a guy with a head for figures who isn't afraid to make up his mind." Collingwood was up for a sequel, saying, "We got to be good friends that night."[47] Television helped turn computer objects into computer *subjects*, a kind of thing that you might have a human-like relationship with. The existence of talking machines like the Voder had already shown that machines could seem to communicate as humans do. Some people looked forward to having electronic brains in the social world of human beings, but others weren't so sure.

The boundaries of agency between people and electronic brains became a source of comedy on 1950s television, helping viewing audiences navigate and negotiate the emergence of high-speed calculation into society. In Western traditions, both comedy and tragedy are understood as means of allowing audiences opportunities to work through the conflicts presented by life's incongruities, especially the lack of control experienced in a world of chance and luck. The philosopher John Morreall explains the contrast:

> Tragedy valorizes serious, emotional engagement with life's problems, even struggles to the death. Along with epic, it is part of the Western heroic tradition that extols ideals, the willingness to fight for them, and honor. The tragic ethos is linked to patriarchy and militarism . . . and valorizes . . . blind obedience, the willingness to kill or die on command, unquestioning loyalty, single-mindedness, resoluteness of purpose, and pride. Comedy, by contrast, embodies an anti-heroic, pragmatic attitude toward life's incongruities. . . . Comedy has mocked the irrationality of militarism and blind respect for authority. Its own methods of handling conflict include deal-making, trickery, getting an enemy drunk, and running away. . . . In place of Warrior Virtues, it extols critical thinking, cleverness, adaptability, and an appreciation of physical pleasures.[48]

Comedy often undermines elitism, providing a voice for the disempowered and the common. UNIVAC shows up in several comedic genres on 1950s television, giving audiences the opportunity to mock computation's oppressive rationalism, its disembodiment, and its increasing position of power at the center of both politics and the economy, while also debating its role in the social world.

One of the most well-known stand-up bits in all of 1950s television was the star-making performance of comedian George Gobel explaining the electronic brain. The bit was part of a two-hour variety special on October 24, 1954, called *Light's Diamond Jubilee,* commemorating the seventy-fifth anniversary of Edison's electric lightbulb. Produced by David O. Selznick and including an impressive lineup of Hollywood stars as well as a message from President Eisenhower, the special received mixed reviews overall, but the one standout was Gobel's twelve-minute monologue. *Billboard* called it "a sheer delight," and the *Washington Post* called it "high comedy at its best."[49] The *New York Times* described it as "a priceless satire on the world of science and its gobbledegook" and highlighted Gobel in its review as "walking off with the major honors. . . . For sheer sustained laughter there has been nothing like it on TV for a long time."[50] *Variety* found the entire show lacking, but called out Gobel's bit as "a completely delightful turn . . . that projected Gobel to the top of the comedy heap."[51] Indeed, he got his own comedy show the next year.

Gobel's monologue followed a narrated video montage of electrical "labor-saving inventions" that were soon to make home life easier, including the "push button magic" of electronically controlled kitchen equipment like freezer-door ice cube dispensers and ultrasonic dishwashers, magnetic taping of television programs from wall-mounted thin television sets, environmental sensors that triggered automatic window closing, and even video baby monitors and pocket telephones (at a time when direct dialing itself was not yet universal). Of course, there was a little bit of Cold War national defense thrown in, but these emerging machines were mostly the first wave of future smart homes. It was the World of Tomorrow 2.0.[52]

This segment ended with images of a mainframe computer, as well as an omniscient narrator who changed tone quite dramatically, from promoter to parodist, with an exaggerated cadence alluding to Remington Rand's UNIVAC advertising: "High on the list of wonders already with us is . . . the electronic brain! This is the great know-it-all of the future. He, or it, is already able to add faster than the entire income tax bureau, translate languages, to tabulate, evaluate, dissertate, innovate, recreate, inoculate, precipitate, but not, as yet, procreate. What, you may ask, is the application of these talents to the life of the common man? For the answer, let us call upon a common man."

The cheeky introduction set up Gobel, known for his homespun style of humor, who began matter-of-factly, "Well, now, I'm a common man all right," to outbursts of laughter from the audience. With a pointing stick in hand, Gobel addressed a frame on the wall that contained a drawn schematic that he identified as "the wiring diagram for the electronic brain," tapping the frame to advance the diagram. The gag was in the fact that Gobel would point to something on the diagram, as if to explain it, and then step away to ad-lib about something having nothing to do with technology before saying, "and that brings us back to the electronic brain." This "common man" obviously knew nothing about computers, but he did know about the absurdities of American life in the 1950s.[53]

Gobel used hyperbole to mock the way that computers were promoted as fast and efficient: "Did you know that it took 700 trained engineers 2,600,000-man hours just to draw this diagram? And they worked in complete secrecy. Not a one of them knew what they were doing." At one point, he reminded the audience that "not one single bit of this whole complex mechanism will do anything if some clown forgets to plug it in." This particular joke, low-hanging fruit though it may have been, shows up as a trope in later television comedy poking fun at computers. The simple fact of their needing electricity to work gives the "common man" a sense of agency over computers' power. Other jokes mocked the drudgeries of daily industrial life that remained outside the realm of calculation, like unfulfilling work and difficult relationships. In joking about the computer in comparison to these human problems, Gobel undermined the idea that the machine is centrally important.

The conclusion of the bit consisted of a series of rhyming lists echoing the narrator's introduction of things the electronic brain could do. Everything in Gobel's lists reflected the mundanity of an everyman's daily life that computation did not apply to. He riffed, "[The electronic brain] can sew stitches, dig ditches, fix britches, tie hitches, and count riches. It can comb hair, shoot bears, cut squares, make chairs, and take dares. . . . And can mix a perfect martini while refereeing a tennis match, which is an important social asset." The activities were all physical, highlighting the obvious embodied differences between people and electronic brains that are actually a useful feature of human beings. By pointing out the things the electronic brain "can" (but actually can't) do, the fool, or "common man," regains control over the authority of the computer. Morreall describes stand-up as a genre like philosophy, in which the comedian asks whether familiar ideas make sense and refuses to defer to authority and tradition. Gobel's self-deprecating "common man" undermines the efficiency claims for electronic brains by stressing the physicality of everyday life, as well as its illogical social contingencies.

Comedy also tested claims about UNIVAC's powers of prediction. It wasn't long before this was put to the ultimate human test: matchmaking. In 1956, *Variety* announced a new gag for the upcoming season of Art Linkletter's *People Are Funny*: "You might as well give up, fellers; automation has invaded the field of romance," teased the blurb. The idea that the computer was an oracle was the boundary to be tested through the electronic brain's matchmaking. Using the machine for such a purpose was not a unique idea; Linkletter's production company was sued by two separate writers claiming that the show had stolen the idea from them.[54] Singles were recruited through newspaper advertisements seeking volunteers who were willing to have the computer "heart machine" play cupid for them.

Ahead of the show's third season in 1956, 4,000 volunteers completed a thirty-two-item questionnaire about themselves and their dream partner that was designed for that purpose by conservative pop psychologist Paul Popenoe. On the season premiere, twenty-one-year-old Virginia Meyer was chosen as the first contestant to be introduced to her UNIVAC-selected "ideal mate."[55] In a format that would later be emulated by *The Dating Game*, Virginia was introduced to three men, one of whom had been identified by UNIVAC. No one on stage was told ahead of time which man UNIVAC had picked, and there was $1,000 in it for Virginia if her pick ended up being the same as UNIVAC's. She picked Ed Hakim, an advertising man, just as the UNIVAC had, and three weeks later, they returned to the show and reported that things were going "just fine" and they had many common interests, including dancing, singing, and hosting parties. The next attempt at UNIVAC matchmaking was less successful. The television critic for *Variety* reported:

> From the so-called scientific data, Univac is supposed to select the proper mate for each of the volunteers. On Saturday's show, the mechanical marriage counselor was at odds with the twenty-three-year-old airline stewardess' understanding and taste. Of the three handsome men on stage, one was selected by the machine. Each time she chose the "wrong" man, when asked which man she preferred and which man she thought the machine selected. She dated the man she liked while the machine (who knows?) had heart failure.[56]

This was not the end of UNIVAC dating, though. The gimmick played on throughout the season. Ads in the back of women's magazines offered computerized matchmaking, as well as computerized horoscopes. The idea of knowing your destiny, either through the mysteries of mathematics or of the stars, had some appeal. At the same time, there was comfort in the fact that the human heart was mysterious and unknowable to the machine.

Of course, not everyone was convinced that the computer was a crystal ball, whether or not they had a sophisticated technical understanding of statistics. In some television comedy, the computer's mathematical nature could not explain human foibles. Human subjectivity was reasserted as individuality was celebrated, even if that individuality was a little quirky. A 1958 episode of the sitcom *How to Marry a Millionaire*, a spin-off of the 1953 Marilyn Monroe/Lauren Bacall film, parodied the *People Are Funny* stunt.[57] Loco, the ditzy character played by Betty Grable in the film and by Barbara Eden in the sitcom, is talked into being matched on live television with her ideal mate, from a pool of 5,000 eligible men, by a computer called the Truthivac. In a parody of the way that computers were always promoted for the speed with which they performed their calculations, the television producer tells a reluctant Loco, "In a fraction of a second, Truthivac can discover incompatibilities which would take nine years of living together to uncover!" Still, Loco is determined to pick a husband herself until it's explained to her that she doesn't actually have to marry the man that Truthivac chooses, and she gets to keep $500. Loco's roommates further encourage her to participate because they disapprove of the man she has been dating and are anxious to have Truthivac tempt her with a more suitable match. The roommates believe in the computer's matchmaking ability, in the reliability of its data. It's the ditzy Loco, the true romantic, who is the skeptic.

Loco finally agrees to the stunt. However, without her knowledge, her boyfriend, Douglas Brock, bribes the computer's operator, Tucky, to have Truthivac match them up. The roommates discover that Brock's data has been added to the set of potential mates, but they are so convinced of his incompatibility with Loco that they aren't concerned; the Truthivac, based on scientific principles, would never be wrong about something so obvious.

On the day of filming, Tucky "throws the switch" (literally, as the computer has a gigantic switch as its main controller) and Truthivac lights up, its tape reels spinning. After a couple of seconds, a single, punched card falls into a tray on the front of the machine. Loco's perfect mate is announced as Brock, who runs in from offstage to embrace her as she squeals with excitement. The set parodies the computer, which is staged as a cartoonish grid of blinking lights that Tucky has to bang with his fist to get working. Its large control switch is reminiscent of both Dr. Frankenstein's lab in the 1931 *Frankenstein* film and the machine floor of Chaplin's *Modern Times* (1936). The effort needed from the room-sized machine to produce a little card with a single name on it also adds a touch of absurdity.

While Loco and Brock are eloping, the roommates discover Brock's bribery scheme and set off to find Loco and tell her the actual truth—as opposed to the Truthivac's truth. Loco is indeed angry about Brock's deception and halts the wedding. "If there's anything I can't stand, it's a fella who will force a man who is running an electronic machine to tell lies!" Loco yells at Brock, to which he replies, "If there's anything I can't stand, it's a girl who needs a machine and an old strudel of a scientist to help her make up her mind!" The computer is explicitly not an agent in the decision making that they each describe. The "old strudel" is the computer's creator, Dr. Millmoss, a caricature of the eccentric physicist, complete with stereotypical German accent and Einstein hair. Loco is seen later at the computer with him as he tries to fix the problem created by manipulating the machine to select Brock, a human comedy of errors that obviously has no technical fix. Eventually, the ditzy Loco, herself, asserts control over the machine when she gets it to work properly by noticing that it needs to be plugged in. Dr. Millmoss's own ridiculousness is apparent when he says of her, "This one has a good scientific mind."

In addition to showing the computer as subject to human corruption, Truthivac "fails" on its own terms when it finally selects the idiotic Mr. Montgomery as Loco's ideal mate. The computer is again being blamed for assumptions that would have been programmed by people, but the source of the comedy is in the fact that an illogical human axiom such as "opposites attract" can best computer prediction.[58] When Loco is introduced to Mr. Montgomery, she is disappointed to find him unattractive and stupid.

Taking the Truthivac's inventor aside, Loco asks if she is really like her match, and Millmoss tells her, "Of course, only on you it looks good!" Unfortunately, sexism (as comedy) gets the last word, but the idea that human attraction is computable has also been ridiculed.

In this episode, the computer is not at all personified. It doesn't communicate except via blinking lights and punched cards, and only when operated by a human being. Vocal stereotypes are used to characterize the human cast—the ditzy blonde with her childlike voice, the eccentric European physicist clueless about the human condition. Here, Loco emerges in control, as she's able to evade attempts by the Truthivac, Brock, and even her roommates to direct her actions. That is not to say that there's no misogynist humor involved, but as with Gobel's "common man," Loco is the ditz/fool who reveals what is true. In showing that the surprises of human relationships are the real riches of human connection, each undermines the authority of the computer that would predict their destiny for someone else's purposes.

There's been an assumption in both the history of computing and in cultural criticism that the general public didn't do much thinking about computers in the mainframe era, as there were relatively few computers and only a small percentage of people had any firsthand experience of them. In contrast, these few examples from 1950s television show that the viewing public had opportunities to consider the meaning of the computer's emergence in political, social, and cultural life. Unlike the barrage of television science fiction that would hit screens in the next decade, many of the televisual representations of computers in the 1950s were of real, or realistic, computers. That they were often described in human terms as autonomous "brains" that communicated their own ideas was an exaggeration that hid the role of humans in the work of computerization, especially in the design of data models and the embedded human biases baked into them. Even if anthropomorphism promoted a misunderstanding that conflated prediction and probability, it also provided opportunities to challenge and negotiate the role of computers in society. Comedy offered a safe way to ask pointed questions about the value of rationality over the uniqueness and uncertainty that are characteristics of being human.

Where computerization caused the most anxiety, though, was in its potential impact on economic life. Very real concerns about automation were also addressed in the media, and computer manufacturers were keen to counter them. Like Remington Rand, IBM got into the computer-manufacturing business in the 1950s and, following Remington Rand's lead, began sponsoring television programming in the late 1950s.[59] This was only one tactic of the company's wide-ranging public relations campaign. As Westinghouse had done with its robots, IBM worked to show computers and the corporations that provided them as key to the success of white, middle-class Americans.

THE INFORMATION MACHINE

In his insightful history of the US office, Nikil Saval explains the zeitgeist among office workers in the 1950s as a combination of postwar anxiety about the conformity that office work required, the bureaucracy of increasing corporatization, and the possibility that suburban prosperity could be yanked away in the midst of Cold War paranoia, if not nuclear annihilation.[60] This scary side of the total corporate environment was captured in some of the most widely read books of the decade: sociological studies like *The Lonely Crowd* (1950) and *White Collar* (1951); *The Hidden Persuaders* (1957), an expose of the public relations world; the management bible *The Organization Man* (1956) and its *Mad* magazine parody; and, of course, *The Man in the Gray Flannel Suit*, the quintessential novel of postwar, suburban ennui, made into a popular movie in 1956 starring Gregory Peck. As Saval summarizes, "Being middle class in America used to mean starting your own business; by 1950 it meant, almost invariably, that you put on a suit and tie and went to work in an office, alongside millions like you."[61]

IBM had the prototypical total corporate environment, with its dress code, militaristic hierarchy, and team-building hymnal. It entered the inaugural *Fortune* 500 list in 1955 at number 61.[62] More than 56,000 people worked for IBM in that year, mostly men, many of whom were subjected to the uniform dress code of dark gray suits, black ties, and starched, collared white shirts.[63] Saval explains, "The uneasiness of a computer company having slotted its employees into a uniform corporate dress, like so many lines of indistinguishable code, was lost on nobody—including the company."[64] He cites a customer brochure from 1955 that read, "We first came into your life when your birth was recorded on a punched card. From then on, many such cards have been compiled," and the text went on to list many examples: education, hospitalizations, major purchases, income tax records, and so forth. IBM may have stood for "international business *machines*," but it was a company in the business of *information*, which for the Information Age meant data processing that compiled a person's lifetime of experiences from punches on cards. The brochure claimed that this was some sort of paternalistic caretaking, but IBM's growing monopolization of data-processing machines did not sit well with everyone, including US federal regulators. IBM needed to control the narrative told about it and its products in the media.

When Thomas Watson, Jr., took over the presidency of IBM from his father in 1952, he made two significant changes: he steered the company to focus on the manufacture and sale of electronic computers, and he reorganized the company's pyramid management hierarchy into a horizontal structure that he believed would be more efficient. Taking seriously the slogan of the Deutscher Werkbund, that "good design is good business," Watson also hired designer and architect Eliot F. Noyes, a protégé of the Bauhaus guru Walter Gropius, to be IBM's new "consultant director of design" and to entirely reinvent its corporate image by redesigning everything "from curtains to computers." They were engaged in what one IBM engineer characterized as a "design battle" to establish control over the public image of the computer.[65] The design historian John Harwood documents this fascinating and understudied relationship between IBM technologies and midcentury modern designers, which exposes the links between US Cold War political ideologies and IBM's public relations campaign to sell the very idea of computers to Americans, decoupling computers from their wartime association with the bomb and with totalitarianism, and reconciling, again, the incompatible ideas of corporate capitalism and democracy in the American political imagination, this time with an Information Age spin that computers were the engines of human creativity.[66]

Noyes had retained Charles Eames as a consultant and commissioned the Eames Studio to create public-facing short films and exhibits that furthered

IBM's need to communicate that the computer was a social good. The first film that they made was *The Information Machine or, Creative Man and the Data Processor*, a ten-minute animated short produced for IBM's exhibit at the 1958 Brussels World Fair.[67] Unlike the machine-centered framing of *Cybernetics* or *Giant Brains*, *The Information Machine* began with artists, defined in the film as people throughout history who were curious and observant, who were "constantly building up stores of information in active memory banks" that they then used to solve problems, "to speculate, and to predict." This expanded the electronic brain metaphor by making memory and information processing analogous human and computer characteristics, but computers still excel at the amount of information they can store and the rate at which they can make sense of it. The calculation that computers did was now "information processing" that created knowledge. The message of the film is summed up in the synopsis written by Eames: "The computer is a tool for handling information that is capable of turning inspiration into fruitful prediction. It offers the promise that there will be room in our thinking efforts for those smallest details that are the basis for man's most rewarding wishes. The computer is an instrument in the service of mankind."[68]

Within the film, this point is made as cartoon figures of the quintessential IBM men—white-collared shirts, skinny black ties, crew-cut hair, and peach-colored skin—happily push buttons on IBM equipment as the narrator states unequivocally, "With the computer, as with any tool, the concept and direction must come from the man." At this point, "information" is introduced along with motion photography of punched cards being shuffled into an IBM computer, making the information/data definition visual, as the narrator explains, "the proper use of [information] can bring a new dignity to mankind." IBM's own history acknowledges the lofty cultural goals of Watson: "When you just want to sell machines, you describe their features and functions. . . . When your ambition isn't just to build information technology, but an information age—that's different."[69] IBM would take the lead in promoting this Information Age around the world and provide much of the infrastructure that enabled it in the decades to come. In the process, IBM would become one of the first corporations to install talking computers, making a key metaphor of computer communication a reality.

3 AM-QUOTE (1964)

In April 1965, *TIME* magazine's cover story announced, "The Computer in Society," illustrated with a large, anthropomorphized mainframe feeding itself punched cards being served up by a female cartoon figure holding a silver platter. Half a dozen men in business attire poured over printouts as the computer's many arms multitasked above them.[1] The accompanying article asked a number of perennial questions about computers: Is the computer a friend or enemy of humans? Will it cause hopeless unemployment by speeding automation? Will it devalue the human brain or happily free it from drudgery? Will it ever learn to think for itself?[2] *TIME* advocated for the information machines.

Citing the requisite experts, the article declared that the computer had more beneficial potential for humanity than any other invention in history. Among its achievements: new horizons in science and medicine, improved techniques for education, greater efficiency in government, reduced costs of production, and lowered barriers to knowledge. The article provided example after example of computers' "superhuman" powers, stressing that they could "run things" more efficiently than people. The article used the verb "cybernate," which now meant to control a process by means of computers. The country's growing dependence on computers wasn't seen to undermine their value, even though the article explained that "if all the computers went on the blink, the country would be practically paralyzed: plants would shut down, finances would be thrown into chaos, most telephones would go dead, and the skies would be left virtually defenseless against enemy attack."[3]

(The reality of this dependence caused a near-panic when the Y2K bug was brought to the public's attention in the late 1990s.)

Nevertheless, it was a fact that computers had taken over the running of key infrastructure across the world. According to *TIME*, the logical next step was to take full advantage of this "dynamic alliance" between people and computers and invite the computers home. What we often call the "personal computer revolution" was still more than a decade away, and it would be 1982 before *TIME* recognized the personal computer as "machine of the year" (in place of its usual "man of the year"), but the magazine was already anticipating the individual relationships that people could have with computers.

The 1965 cover story suggested that it was people's responsibility to tame computers, machines it described as living in "families" and which needed to "get out and see life" in society, lest they end up "mumbling to [themselves] in the air-conditioned seclusion of [their] . . . data-processing room."[4] While the "problem with having a machine for a buddy" in 1965 was that "it does not make for a very good conversationalist," the article assured readers that "the day is clearly coming when most computers will be able to talk back."[5] They highlighted such a talking computer installed at the New York Stock Exchange (NYSE) that was providing instant stock quotes over a special telephone.

The talking computer at the stock exchange was one of the first real-world examples of voice synthesis deployed as a human-computer interface. Voice synthesis research struggled for funding in the 1960s when the space race was the US federal priority, but the achievements in computer miniaturization needed for the *Apollo* program had impacts for science and technology beyond the National Aeronautics and Space Administration (NASA). The most important thing to happen to voice synthesis in the 1960s was the development of the transistorized minicomputer. As computer historian Paul Ceruzzi has described, "Digital computers began the decade of the 1960s with a tentative foothold; they ended with an entrenched position in many business, accounting, and government operations. They were also now found in a host of new applications."[6] The minicomputer didn't democratize computing for the masses, but it had a bit of this effect among researchers. Those who had been at the back of the line for sharing mainframe resources were acquiring minicomputers that they could dedicate to their own research. In the speech community, this research was largely related to signal processing. From the human-operated analog Voder through the hardware embodiments of electronic vocal tracts (EVTs), researchers could now transition their models to digital computers and manipulate them with programming. Consequently, the number of research labs working on voice synthesis increased, and the first businesses hoping to sell voice synthesis as a service also emerged.

This chapter explains some of the key research developments in the 1960s that would finally lead to consumer applications of voice synthesis in the 1970s. Some of these research activities also had some surprising influence on popular culture, especially at Bell Labs, where the cultural milieu of the 1960s spawned the Experiments in Art and Technology (EAT) collective, which brought New York City artists and Bell engineers together in avant-garde collaboration. In the meantime, IBM and other companies were working to make voice synthesis an interface for customers needing up-to-the-minute access to information processed by computer including the finance industry, which was among the first to implement talking computers. With popular attention turned toward the accelerating technological developments shaping work, facilitating the race to the Moon, and bringing the Vietnam War right into people's living rooms, science fiction stories on television and at the movies gained a wide audience, but this chapter shows how portrayals of talking computers also made their way into more down-to-earth genres. Even Disney movies engaged the cultural conversation about the important differences existing between people and talking machines.

DIGITAL SIGNAL PROCESSING AND SINGING COMPUTERS

Digital signal processing (DSP) research took off in the 1960s with the availability of minicomputers that were ten times less expensive than their mainframe forebears. In some respects, DSP research had also driven minicomputer development. At the beginning of the Great Depression, a small company was formed that provided what was essentially a signal-processing service for the oil and gas industry. Called "reflection seismography," the process used small explosions of dynamite to trigger vibrations, or wave energy, through the Earth that were then visually recorded. These seismographs were used to identify where oil and gas reservoirs were located. In the same way that Bell Labs was looking to expand telephony by researching voice signals in the 1940s, this geophysics company was looking for ways to expand the efficiency and effectiveness of their seismograph technology. Having expanded to military applications during the war, in 1951 the company changed its name to Texas Instruments (TI).

TI's leadership immediately saw the potential benefits of the transistor, debuted by Bell Labs in 1948, and purchased a license to manufacture the technology in 1951 (as did General Electric, RCA, Westinghouse, IBM, and several other companies). In 1953, TI researchers built their first special-purpose computer for seismographic analysis. Unlike some of their competitors, TI's semiconductor research and manufacturing started from scratch, but together with the drive to improve their special-purpose computers, and a key scientist, Gordon Teal, poached from Bell Labs, they surprised the industry with the development of a silicon transistor in 1954 and with Jack Kilby's integrated circuit (IC) in 1959.[7] TI built one of the first computers to use ICs, and the US Air Force's need for miniaturized missile electronics helped develop the microcontroller technology that would eventually enable the personal electronics and personal computer industries.

Circuits relayed signals, and the development of digital methods made it possible for almost anything to be converted to a binary signal. One advantage of DSP is the greater ability to manipulate a signal, either to enhance features of it, extract information from a noisy signal, analyze a waveform into its component frequencies (as for purposes of synthesis), or to encrypt or compress the information in the signal for transmission. The historian Frederik Nebeker lists the signal processes developed by engineers in the 1960s as "filtering, coding, estimating, detecting, analyzing, recognizing, synthesizing, recording, and reproducing."[8] No matter whether the signal is for speech synthesis or seismic analysis, the general sequence is the same: an analog waveform passes through a filter that removes frequency components above a certain frequency to ensure that low-frequency aliases of these high-frequency components aren't sampled. After this antialiasing filter, the signal is sampled at particular points in time and these amplitudes (heights of the signal waveform) are quantized.[9] This is the analog-to-digital conversion process, whereby the continuous-signal waveform is converted to discrete sets of values that *represent* the waveform. In other words, the sampled amplitudes, once measured, are converted to binary, discrete numbers. Once the digital-processing task has been completed, a reversal of these steps transforms the updated digital sequence back to an analog waveform. As has been described, one of the earliest applications of DSP was speech coding, which could be accomplished because of the relatively low bandwidth of speech signals compared to, say, optical signals. "It is ironic that digital communications, long a stepchild of the telephone industry, now promises to ease the problems of overcrowded voice transmission facilities," noted the Institute of Electrical and Electronics Engineers (IEEE) in 1973, possibly having Claude Shannon's 1948 coauthored paper on pulse code modulation with his colleagues Bernard Oliver and John Pierce in mind with the term "stepchild."[10]

The objective of speech coding is to represent speech with the fewest bits necessary for it to still be perceived as speech, compression required to scale telephony, but also, it would turn out, for consumer-grade speech synthesis. James Flanagan, head of the Bell Labs acoustic research department in the 1960s, explained how Bell researchers first incorporated digital processing into their speech synthesis research:

> We found that you could get useful solutions for just about any signal process that you could turn into difference equations. That's the way we did most of the speech synthesis, with vocal tract simulation by differential equations, then solving them simultaneously. Not in real time, obviously. This early speech-processing work required sampled data, and the understanding of sampled data signals. But it also needed all kinds of filtering and spectral analysis. I think the whole area of digital signal processing . . . was driven by the speech processing community.[11]

As a signal, speech was found to exhibit considerable redundancy due to both the physical mechanism of the human vocal tract and the structure of language. In addition, Bell's research on hearing led them to understand that the human ability to perceive speech and other sounds is constrained in both dynamic range and bandwidth. Flanagan, again, likened the human ear to "a Fourier spectrum analyzer or filter bank," which itself participated in a compression/perception process.[12] With all this natural redundancy in speaking and hearing, digitalization of speech seemed an achievable goal.

The active speech research area at Bell Labs remained a leader. In oral history, the electrical engineer James Kaiser reminisced that when he joined the speech research group at Bell Labs in 1960, "a change in the means of doing research in the speech area, in the coding area, was under way. Instead of the 'old way,' which was to test an idea about a new way to do things by designing the electronics that embodied that idea and then running tests on that physical embodiment, we were starting to simulate the system . . . on the general-purpose digital computer. . . . It was much faster and more versatile."[13] EVTs were being superseded by computer programs.

Of course, the human voice is not "born digital," and there was a lot of work to be done to find efficient ways to get the vocal signal into the computer to be able to work with it. Kaiser explained how he made the transition from analog electronics to digital simulations:

> Right at the time that I arrived there was a marvelous piece of new software that had been developed by John Kelly, Vic Vyssotsky, and Carol Lochbaum called the BLODI Compiler—the Block Diagram Compiler for simulating digital systems. So if you could draw a block diagram for your system, you could usually describe those individual blocks with simple instructions in the Block Diagram Compiler. That would enable you to compile all the interconnections for the blocks that you had and produce a program into which you could just enter your input and get the results out. Now Hank McDonald . . . had built. . . . basically an analog-to-digital converter. It would write a digital tape at about 40 kilo samples a second, if I remember correctly. Then you could run it back the other way, digital-to-analog. So if you were going to process speech, you'd talk in a microphone and [McDonald's] gear would digitize a tape. You would then take that tape and a deck of punched cards as input to the BLODI Compiler—one card per block in your block diagram—over to the general-purpose machine. You'd submit the deck with the tape, and back would come another tape. You went back to [McDonald's] gear and listened to the result. It was really marvelous.[14]

Kelly and Lochbaum themselves were working on converting the hardware voice synthesizers of the 1950s to software, specifically programs that they could run on one of Bell's IBM 7090 computers. Their goal was to produce synthesized speech from "input consisting only of the names of phonemes and a minimum of pitch and timing information."[15] The output was a waveform on digital tape that could then be converted to analog. They presented a paper in 1962 in which they described the speech quality they achieved as "far from satisfactory" and working with the IBM 7090 as "laborious," but they expressed optimism about the articulation models on which they had based their experiments, especially the vocal tract analog.[16]

Kelly and Lochbaum's experiments also became a footnote in popular culture, thanks to an interest that Shannon and Pierce had in making music with computers. Shannon was an avid clarinetist who haunted the jazz clubs of 1950s Manhattan. Pierce could hardly play a note, but music was a lifelong passion. Together, they had tried to figure out the information rate of music. During a classical concert in 1957, Pierce suggested to Bell Labs engineer Max Mathews that he take some time to write a program for making music with a computer. In Mathews's recollection, "Pierce said, 'You get sound out of a computer now, you get numbers out of sound, if you write a different program, maybe you can get computers to make music.'"[17] They justified the project to management on the basis that it would provide insights relevant to voice synthesis. This is how Kelly, Lochbaum, and Mathews ended up programming the IBM 7090 to sing "Bicycle Built for Two (Daisy Bell)." No one bothered to document why they chose "Daisy Bell" for their experiment except that Mathews once said that it contained a lot of long vowel sounds that were easier to synthesize than other vocal sounds.[18] In 1962, Pierce and Mathews produced an entire LP of computer-generated music on Decca Records titled *Music from Mathematics*. Pierce played the recording for eminent British science fiction writer Arthur C. Clarke during a visit that Clarke made to Bell Labs in that year, and it became the inspiration for the HAL 9000 computer's "death" scene in *2001: A Space Odyssey*, on which Clarke collaborated with auteur filmmaker Stanley Kubrick.[19]

Many artists collaborated with Bell Labs engineers in this era. Bell Labs engineers Billy Klüver and Fred Waldhauer launched the nonprofit EAT collective in 1967 with artists Julie Martin, Robert Rauschenberg, and Robert Whitman. In response to some of the public ambivalence about emerging technology and the enthusiasm in the 1960s for participatory performance art "happenings," the group's aim was to "eliminate the separation of the individual from technological change and expand and enrich technology to give the individual variety, pleasure, and avenues for exploration and involvement in contemporary life."[20]

Max Mathews's work in computer software for electronic music was also pioneering.[21] He often made presentations and worked with university music programs and labs in the Northeast. A growing number of composers and musicians gravitated toward electronic music in the 1960s. Don Buchla formed his electronic music equipment business in 1962, and Robert Moog's analog synthesizer debuted in 1964. Electronic music did not remain avant-garde; as the musicologist Timothy D. Taylor has documented, electronic music was "subliminally" fed to the listening public through radio and television commercials in the 1960s. Electronic music, the stock-in-trade of the strange worlds of science fiction cinema, was domesticated for selling products to prosperous post–World War II suburban households. Afforded by developments like Moog's synthesizers, commercial composers set about giving inanimate commodities, from coffee pots to vacuum cleaners, "friendly" voices, bringing them to life through electronic sounds.[22] By the end of the decade, Wendy Carlos's album *Switched on Bach*, containing forty minutes of Johann Sebastian Bach's compositions painstakingly arranged and played on Carlos's customized Moog synthesizers, became a surprise hit.[23] The album was certified gold (over 1 million copies sold) by the Recording Industry Association of America (RIAA) within a year of its release and won three Grammy Awards, including Best Classical Album and Best Classical Performance.[24] Carlos had been affiliated with the Columbia-Princeton Electronic Music Center, where Mathews was well known. She went on to compose the soundtrack for Stanley Kubrick's *A Clockwork Orange* (1971), among many other projects.

The composer Charles Dodge was a graduate student also involved with the Columbia-Princeton Electronic Music Center in the second half of the decade. He reminisced that he drove "something like 25,000 miles between Columbia, Princeton, and Bell Labs" in those years, even though the longest leg of that trip is only about thirty-five miles.[25] Dodge is known for his experimental compositions with synthesized voice; he actually gained visitor status to work with the speech research team at Bell Labs in the early 1970s. In addition to Mathews and Pierce, Joseph Olive had joined the group as an engineer and also had a toe in the music world. Dodge used Olive's speech synthesis software to experiment with synthesized vocal music on *Speech Songs* (1973), but he concluded:

> No, computer speech isn't really very good for music. Of course, the fault was in the preconception of what constituted vocal music. I had this notion that it was a bel canto voice that did certain things. The pitched speech was fascinating and there was a contradiction that took me about a year to work out, so that by the summer of 1972 I got back out to Bell Labs with these Mark Strand poems convinced that I could really make music with it. Then I didn't try to make vocal music, I just tried to make speech music. . . . I had a Guggenheim Fellowship in 1972–1973 that gave me the space to pursue that.[26]

In the time that Dodge worked on *Speech Songs*, Bell Labs engineers made enough progress in speech synthesis that he reflected that the first song sounded "very crude" compared to the fourth, which he described as "fairly natural" thanks to breakthroughs in speech coding by the engineer Bishnu Atal.[27]

Researchers at Bell Labs had been working on a system that implemented pulse code modulation (PCM) since 1955 and finally brought it online in 1962, the world's first common carrier digital communications system. This line sampled a voice waveform at 8,000 times per second and encoded it at 64,000 bits per second. The digital system spread with the availability of ICs, although it took decades for the entire telephone system to be converted. In 1967, the Bell researchers Bishnu Atal and Manfred Schroeder experimented with adaptive predictive coding, a method of transmitting only certain parameters of the waveform, those features studied through the earlier development of the EVTs. This reduced the number of bits per second needed to transmit the voice from 64 to 4.8 kbps, but the quality was only fair, and there was a trade-off in the increased amount of computation needed at the transmitter and the receiver.

Atal kept at it and came up with a better algorithm: linear predictive coding (LPC), a statistical operation for estimating the future values of the shape of the waveform based on previous samples. Atal traced his ideas about prediction back to a seminar in information theory that he attended in graduate school.[28] From the work of Shannon (1948), Wiener (1949), and MIT electrical engineering professor Peter Elias (1955), Atal became interested in the theory of predictive coding that explains how, when both a transmitter and a receiver store the past values of a signal, a present signal can be assumed from transmitting only the prediction error, or the difference between the signal and its predicted value (basically, the "noise" rather than the signal).[29] At the receiver, the prediction error is then added to the predicted value to recover the signal.[30] This required smaller samples for analysis and proved to be better for synthesizing rapid changes in pitch, such as those characteristic of higher voices. This was a change from some of the assumptions in Homer Dudley's analogy of the "carrier nature" of speech, which specified that speech was carried by low-frequency modulation signals that corresponded to the motion of the vocal organs and compression was achieved by extracting these signals.[31] Dudley's ideas required the buzz of periodic voiced speech and the hiss of nonperiodic unvoiced speech, but predictive coding suggested that both could be re-created from noise alone. Linear prediction of a signal was developed by Norbert Wiener and is also fundamental to Claude Shannon's information theory, and Atal's LPC is an algorithmic descendant of their statistical proofs. In fact, mathematically equivalent signal-processing algorithms were proposed independently at around the same time by researchers working on telephony in Japan and on geophysical signal processing in the US.[32]

Even though Atal was writing his PhD thesis on speaker recognition, he decided to pursue his ideas about predictive coding and speech compression as "a side project."[33] After an intriguing presentation at the 1967 IEEE International Conference on Speech Communication, Atal reported that several other research groups started contributing research toward improving LPC for speech coding. Atal's first successful trials required two predictors, one at an interval of about 50 ms, or the time of a single pitch interval, and another at a tiny interval of 1 ms. LPC required a lot of computations, but it worked effectively at a lower bit rate than PCM. Even though the speech quality wasn't good enough for commercial telephony, it worked for toys and industrial applications, as we will see in chapter 4. One researcher working in DSP at MIT related the difference that LPC made to furthering speech research, explaining, "One of the reasons we had been doing radar work [in the 1960s] was that the funding for speech work had dried up. . . . All of a sudden LPC came along, just another bombshell. . . . LPC caused the funding switch to open again, and we got money, and we were able to work on stuff."[34] Both waveform synthesis and LPC synthesis left the lab in the 1970s to be implemented in consumer electronics available to the general public, but even in the 1960s, developments in voice synthesis brought its attention to engineers thinking about the barriers that people experienced interacting with computers, ideas that contributed to the emerging field of man-machine communication.[35]

INVENTING AM-QUOTE

A key marker of the shift from an industrial capitalism to an informational one has been the financialization of the US economy.[36] Often associated with an extremist laissez-faire economic ideology whose proponents succeeded in affecting US policy in the late twentieth century to decrease New Deal–era regulations related to capital, financialization has also relied on the use of computational information technologies to control and manipulate capital as *information*.

Computation has always been constitutive of capital, but electronic computing increased the speed and scale of computational ability far beyond the constraints of human bookkeeping, with concomitant explosions in the scale of financialization itself, as well as accelerating what was perceived as up-to-the-minute information access. Voice synthesis played only a bit part in the initial computerization of US financial institutions, but its deployment foreshadowed how voice would be used as a computer interface in the future, driven by assumptions that voice would be the easiest, user-friendly way to integrate people into human-computer systems of information circulation. In technical terms, this use of voice synthesis leapfrogged some of the ongoing speech research in favor of a concatenative system that will be described next. In political-economic terms, we see voice synthesis used primarily to streamline the sharing of computed information with specific human agents, automating what was previously a human-centered communication chain and consolidating data control at a corporate level. Finally, in cultural terms, talking computers became a real phenomenon, and the way that they were discussed in the press provides some insight into the ongoing cultural negotiation of the role of computers in society.

In 1961, US president John F. Kennedy named as Handicapped American of the Year the inventor Emik Avakian, who worked at Teleregister Corporation of Stamford, Connecticut, as "manager of man-machine interface development," certainly one of the first persons to hold such a position.[37] Avakian, who was born with cerebral palsy, had come to the attention of the press in 1952 when, as a twenty-eight-year-old electronics consultant for IBM, he invented a machine interface that allowed him to control an IBM typewriter using his breath. *Life* magazine published a picture of Avakian in front of a microphone attached to an analog audio control box next to the typewriter and a computer, taking up about one-third of the desktop, which the article identified as "an ingenious junior-size electronic brain."[38] Avakian had worked out a four-digit code using the numbers 1–4 for each uppercase and lowercase letter, the numerals, and special characters on the keyboard, plus codes for backspace, tab, carriage return, and space. The numbers in the code corresponded to a breath blown into one of four microphones, so the lowercase "a," with a code of 3344, required two breaths into microphone number 3 and two into microphone number 4. Avakian was literally converting his body's breath into digital information, much as telegraphy had done with the motion of a human finger. A June 1953 profile in *Mechanix Illustrated* quoted Avakian explaining the human body using the cybernetic metaphor: "A man's body is a vast intercom system serviced by a central transmitter, the brain. Palsy victims such as I suffer a breakdown of communications."[39] In the early 1960s, he helped create a voice interface for accessing stock price information from computers at the American Stock Exchange (ASE).

The system was called Am-Quote. It allowed a stockbroker to dial on the telephone a code number associated with a specific stock and hear a synthesized voice give a real-time report about that stock, all generated by computer. *Life* magazine had again profiled Avakian in 1962 and mentioned this new invention of his in the caption of a photograph that pictured Avakian in his wheelchair as seen through a rack of wiring, as if he were quite literally wired into the system. Am-Quote came online on May 11, 1964, and was expected to eventually serve 500 brokerage offices in the New York area. Above the fold on the front page of the *New York Times* financial section, the headline was "Talking Computer Quotes Stock." The lead reveals an ongoing need to clarify for the public the benefits, and the limitations, of computing: "Am-Quote provides faster, fuller, and more accurate information to stock brokers and thence to their customers." The article also reassured readers that "[the] electronic marvel won't make any predictions about the stock market."[40] In chapter 2, we saw how the computer as crystal ball developed as a popular media trope; and here, press coverage clarifies that "prediction," and with it the possibility of corruption, are *not* facilitated by the computer and its voice.

Am-Quote's voice interface could state the stock ticker symbol, bid, asked, last scale, net change, present volume, and the open, high, and low prices for 1,100 issues. This required the machine to be able to make the sounds for sixty separate letters, numbers, and words. This was accomplished by recording a human saying each required phoneme and storing these recordings in the system's memory. A program organized the combination of recordings needed to speak the stock data, a process called "concatenative synthesis" because the individual phoneme recordings are joined together, or concatenated, to form a new message. Memory constraints required that the recordings be short, and compression needs meant that the original, human voice needed to be as unexpressive as possible. One of Avakian's colleagues, Walter Jennison, actually ended up being the "voice" of Am-Quote. Developers had auditioned some leading radio announcers, but their exaggerated inflections weren't going to result in understandable speech samples once compressed for storage in the system. Jennison didn't even use his natural voice when recording the phonemes. He concentrated on maintaining a monotone.

Am-Quote was folded into an existing flow of information and communication. As the *Times* explained, quotes and sales were relayed from trading posts on the exchange floor to a transmission center in the same building, and from there the ticker signals were transmitted "to the computer at the TeleCenter which enters the data on a storage drum," an explanation that removes human beings from this part of the system, at least rhetorically. The broker becomes the only human agent in the system as the explanation continues: "When a broker dials for a quote, the signal is routed to a query-reply drum that selects the proper information and converts the electronic signal for an audio response to the broker." Am-Quote allowed for a significant number of calls to be answered—it could accommodate 200 separate phone lines at a time—and the *Times* article concluded with a description of what Am-Quote allowed the system to do away with: "Heretofore, the American exchange's telephone quotation service consisted of women operators who sat in a big room lined with chalk boards on the building's seventh floor. They endlessly recited stock prices to the brokers. Except for the telephones, the system was entirely human—prices posted by hand on the boards and repeated unemotionally by operators wearing headsets and clicking telephone keys."[41] The journalist made no comment about this change in process, but he did note that it did not come with a price reduction for subscribing brokers, who would still pay $100 a month for a single line.

Before computer systems, the communication of NYSE floor information involved a network of human beings shuffling paper updates through pneumatic tubes, with a lag of several minutes between the time that a transaction was initiated and when it showed up on the ticker tape. The science and technology historian Devin Kennedy has documented how computerization of the stock exchanges reflected managerial efforts to consolidate control over the clerical labor needed to operate financial markets, as well as influence regulatory efforts to favor corporate structures. In discussing the Market Data System, an amalgam of computers brought online to automate the NYSE in the 1960s, Kennedy notes that regulators and financial managers shared anxieties about "the human intermediaries of financial market data: the teams of clerks, typists and reporters who recorded and managed information about trades."[42] These were among the first parts of the process, then, to be targets for automation. Although regulators also pressed for technological surveillance of speculators and floor traders, Kennedy finds evidence for "asymmetries in the application and uptake of computer systems" that "reflect the distribution of power within market institutions, especially between professional members of the exchange and clerical staff," or between *employees* of the NYSE, with less power, and *members* of the NYSE, with much more of it.[43]

The benefits of automation were reaped by brokers whose firms owned portions of the stock exchanges as members. Kennedy explains the separation in this way: "For the NYSE, the pursuit of efficient electronic markets was tied up with a mistrust of the human work of intermediation. But it was not financial intermediation (the work of brokers on behalf of clients) that concerned them, but *informational mediation*: the work of human clerks, typists, and reporters in the faithful representation and management of market data."[44] The anxiety of the owner class about informational mediation, especially by wage employees, was not limited to the financial sector, as we've seen that it was also expressed by Vannevar Bush when he worried about "girl" transcriptionists, as well as in efforts to automate directory assistance in the Bell telephone system, thereby doing away with the "girl operators" who might be listening in.

Am-Quote was also part of a system to automate informational mediation in the ASE, resulting in similar restructuring of information mediation, even though it was described as facilitating information access. While the use of the voice as an interface was helped along under the guise of "increased access," and the technology itself, outside its context of use in the stock exchanges, might have been able to facilitate that goal, its actual deployment helped consolidate financial power to those brokerages and their brokers who were allowed, and could pay for, this limited access. As Kennedy argues, "information technologies not only undergird social and epistemic orders of the market, but . . . are also artifacts of historical struggles and indices of paths cut short."[45] Accuracy and speed were the values through which Am-Quote claimed its place in the financial information system, and it certainly increased both, but computation itself was driving the need for increased efficiency, as transactions could be computed at scales beyond human ability, accelerating what constituted up-to-the-minute access to data moving through the system.

Kennedy also argues that specific arrangements of financial data infrastructure in the US have been influenced by institutions with both access to computer technology *and* the power and ability to shape technology, practice, and regulation at specific historical points. This also plays out in the example of Am-Quote. The *Times* reported that Avakian and others had filed a patent for Am-Quote, but they weren't the only team working on this specific application of voice synthesis and computing. At IBM, engineers in the Data Systems Division development lab were also working on a Voice Answer Back (VAB) device that would arrive at the NYSE in 1965, some months after Am-Quote came online at the ASE, but it had the extraordinary advantage of IBM's corporate power and computing market share behind it. In fact, only a week after the *Times* article about Am-Quote, they published a follow-up that highlighted all the expected computer innovations coming to the NYSE under the banner of the Market Data System, which would "dramatically" switch "the leadership in electronic wizardry" on Wall Street. Still unnamed at the time the report came out, the new system was promised to come online with "both a machine that reads and a computer that talks," leased from IBM, whose stock was traded on the NYSE.[46]

IBM had been working with the NYSE on automation for some time. As Kennedy documents, "human fallibility and labor speed in the mediation of data had long formed the background of computerization efforts at the NYSE," including a series of studies by Bell Labs and IBM in 1951–1952 that looked at labor-intensive information services and recommended the automation of the Quotation Bureau, essentially the same telephone service described at the ASE by the *New York Times*.[47] In a perpetual technical irony, automating some humans out of a computational information system required simplifying access to computer-generated data for others. A 1962 IBM technical report about the development of a Computer Voice Output Device began with a summary of this "problem": "The ever-widening scope of computer system applications has emphasized the man-machine interface. The fact that the language of the computer is incompatible with the language of man is becoming more evident and more troublesome, particularly as real-time systems are applied in new and diverse areas."[48]

The desire to remove some classes of people from the system required that the system become easier for people to use, overall, but the choice about who gets removed is disguised by framing computers as agents in their own right. The IBM technical report brings the computer metaphorically to life when it explains, "This Voice Answer Back device can create output messages under computer control, either as a result of input requests received by the computer, or simply because the computer has reached a point where it wishes to 'speak.'"

As the Am-Quote system did, IBM's VAB device stored analog voice recordings on a magnetic drum memory. The drum of this first VAB revolved at 600 ms that were divided into two 300-ms tracks in which the recordings had to fit precisely. This meant that the phonemes as recorded were often sped up or slowed down to fit within the 300 ms.[49] If a longer word was necessary, it could be split across the two tracks and fill the 600-ms revolution. As the report explained: "Speech varies from 300 to 800 milliseconds per word, and is made up of sequential damped oscillations called 'pitch periods.' These . . . vary among speakers, and are normally shorter in female than male voices . . . and vary slightly for a single speaker. Techniques were developed for both expanding and compressing words to fit chosen time slots, while retaining intelligibility and acceptable naturalness."[50] While stated as given facts about all voices from all human bodies, these measurements are averages for standard American English as spoken by average white, and usually male, bodies. As with Am-Quote, a monotone voice in a middle pitch range was needed for recordings that could be shifted and compressed as required and still be recognized as English phonemes.

The *New Yorker* magazine published a blurb in its "Talk of the Town" section about IBM engineer Robert Rew, whose "readily compatible" and "readily understandable" Midwestern accent became the "voice" of the computer, which had replaced "a covey of girls who sat at a switchboard on the fifteenth floor."[51] In describing how his voice was used for the VAB, Rew grants the computer its own agency, and in the process normalizes the bias of computational systems toward what is most easily measurable and computable. "The computer doesn't care about your accent, although its program may be somewhat picky," he explained. The *New Yorker* writer dialed up the numbers for a few stocks to "hear what the mechanized man, or the humanized machine, sounded like." The verdict was that "his, or its . . . voice comes across in a jerky yet methodical way that sounded to us exactly like the voices of robots in movies made when real talking robots were still a dream. That is to say, nature is apparently up to its old trick of imitating art, and computers are being taught to talk as computers ought to talk."

Of course, as we've seen, the electronic artifacts of voice synthesis such as the Voder's were already in the public's ears when characters like Robby the Robot hit screens beginning in the 1950s. Filmmakers were taking inspiration from technological developments, at least where audiovisual aspects were concerned. The *New Yorker* writer's conflation of "mechanized man or humanized machine," the fact that the writer no longer seemed interested in which was which, suggests an unconscious acceptance of the cybernetic metaphor. Although meant to be quippy, the idea that "computers are being taught to talk as computers ought to talk" demonstrates how quickly the inevitability of talking computers had become common sense even before they emerged.

Getting rid of the electronic sonic artifacts in voice synthesis was an ongoing project of the speech research community. Am-Quote and VAB, and the computers that ran them, had significant limitations that made it difficult to put these devices to additional uses. They worked well enough in this context because the output to be spoken—stock information—consisted of a limited number of mostly single words (like "one" or "two") that needed to be given in order but didn't need to be joined together in full sentences, which would have required more sophisticated manipulations for articulation and inflection than voice synthesis could produce at this time. The speech research community was invested in solving the problem of creating voice synthesis with more linguistic flexibility, as implied in the standard metric of "intelligible," but was still years away from achieving this. The ASE Am-Quote and NYSE VAB systems were fairly short-lived, as further computer automation included printed quote information by teletype that was quicker to obtain than the spoken data. Even so, a voice interface for computer output remained a typical frame when the media reported on these technologies. "Talking to your computer" became a metaphor for data transmission, with computers "talking" to each other as well as to humans.

In spite of the more affordable and flexible teletype services for data transmission, IBM developed voice output for its modular System/360 family of mainframe computers in production between 1965 and 1978.[52] Other companies also developed similar products, touting the benefits of computer voice interaction. In 1971, the *New York Times* reported that there were eleven companies producing such equipment and quoted the chairman of one of them, Instrument Systems Corporation, as follows: "Companies put information into a computer, but how often do they really get out any useful information? Honestly, I believe that many computer rooms are kept busy turning out reports that are almost never used."[53] It would not be the last time that a company would underestimate the complexity of voice interaction in their excitement to make money. This chairman made it seem as if the storage of 2,000 audio files that constituted every phoneme combination in English seamlessly resulted in the spontaneous generation of any sentence, discounting completely the complex cultural and biological processes involved in turning those letter-sound combinations into meaningful communication. (In addition, he was exaggerating the degree to which other kinds of computer output were ineffective.)

The sounds emitted by these programs were difficult to understand as speech, they lacked the coarticulation and inflections needed for understandable English, and recording and programming custom vocabularies were difficult and expensive to do. Talking computers were an existent, but significantly limited, reality. These issues would follow voice synthesis into the 1970s as significant problems for speech researchers and engineers to solve, but developers on the leading edge were now touting the computer interface as the main goal of voice synthesis research. In more and more professional contexts, increasing numbers of people needed to interact with information processed and managed by computers, and speech, which required no instruction to use, was seen as a "natural" way to make that possible.

COMPUTERS IN TENNIS SHOES

The ability to make speech sounds is only part of what is needed to generate voice communication, but it was a step in the right direction. In the 1960s, voice synthesis research was largely focused on text-to-speech applications—translating letters, as symbols, into their speech sound equivalents. The ability to input a speech wave and have the computer translate it into symbols, or speech recognition, was in its infancy at this time, but it did seem like common sense for voice-as-interface to work both ways: people should be able to talk back to their computers *and* be understood. Research under the banner of "artificial intelligence" seemed to promise this future—a future that science fiction and other popular culture representations of computers were already portraying.

Natural-language processing (NLP) is at the heart of "artificial intelligence (AI)"—a phrase coined by researchers who organized a workshop at Dartmouth in 1956 about implementing symbolic logic with computers. They adjusted the framing of their interests to avoid being so closely associated with the wartime cyberneticists and their "robot brains." Some of these AI researchers had focused instead on machine translation, especially of Russian-language scientific papers, an indication of the Cold War context of early AI research. One of the Dartmouth organizers was IBM's manager of information research, Nathaniel Rochester. At the 1958 Brussel's World's Fair, the IBM pavilion demonstrated a Random Access Method of Accounting and Control (RAMAC) disk drive that was described as "knowing" ten languages, including Russian and Interlingua, the universal scientific language developed in the World War II era.[54] The Dartmouth group was committed to the assumption that intelligence consisted of the ability to manipulate symbols and "every aspect of learning or any other feature of intelligence can in principle be so precisely described that a machine can be made to simulate it."[55] Information theory had already turned language into data. Harkening back to Alan Turing's famous "imitation game" test of machine intelligence, AI research would see how far processing language statistically could be taken.[56]

Language use was a synecdoche for computer sentience in popular culture, a projection of human cognitive ability that helped audiences explore the boundaries between machines and people, but also helped provide a lot of power to the metaphor of machine intelligence. With the Cold War and the space race as the backdrop, cultural representations of talking computers ran the gamut of audiences' reactions to this future, from fearing computing's power, to defending human beings' essential nature, to mocking the entire enterprise. Although the calmly menacing monotone of HAL 9000, the computer running the *Discovery One* spaceship in *2001: A Space Odyssey*, is the quintessential representation of a talking computer for many computer enthusiasts, plenty of other examples show computing in a domestic context, where human sociality is negotiated through talking computers and other vocal machines.

For example, *The Jetsons* (1962) is fondly remembered for its depictions of future technology, like the family's flying car and devoted robot maid Rosie, but it is essentially an animated family sitcom, not that different from the live-action suburban sitcoms that the late 1950s are remembered for. In the early 1960s, *The Jetsons* represented the white, middle-class nuclear family, albeit one transported into a Space Age future. It was created in hopes of capitalizing on the popularity of *The Flintstones*, also an animated family sitcom, this one set in a facetious "Stone Age" past that never existed in the same way *The Jetsons'* Orbit City Space Age future was never meant as a serious prediction.[57]

In parallel to the Flintstones' household appliances, which are usually powered by animals prone to squawking and complaining about their plight, the Jetsons are surrounded by beeping and talking machines, like a male-voiced "laundrojet" that can instantly wash, fold, and mend the family's clothing, but also stutters when it tells a scolding Jane that it hates to sew on "b-b-buttons." The family at home has Rosie, and at work, George Jetson has the talking computer RUDI, his sidekick and Pinochle opponent. RUDI (for "Referential Universal Digital Indexer," a comedic mashup of early computer terminology) is drawn as a mainframe, with lots of blinking lights, switches, tape reels, and meters, as well as a control desk wrapping around the inside of a saucerlike room. Played by the voice actor Don Messick, who also did the voice of Astro, the Jetsons' dog, some of the series' robots, and other characters, RUDI speaks in short phrases, usually in mocking response to something that George has said, and its wobbly voice is visualized by a red waveform on the front of the computer, a developing visual trope for nonandroid talking machines. When George uses the knobs and buttons on the computer to calculate a problem, the answer gets printed on a card rather than spoken by RUDI. It's as if RUDI the coworker and RUDI the computer are two separate beings in a single machine body. These talking appliances are an unexceptional part of mundane daily experience for the Jetsons. They get frequently upgraded and are status symbols of suburban material life. (Rosie is actually a used robot maid that Jane Jetson acquires because she cannot afford the latest model.)

Although the most vocal appliances, like Rosie and RUDI, have human personalities and serve as companions to humans, many of the other talking devices are every bit as much of a nuisance as Wilma Flintstone's sometimes persnickety animal appliances. They are individuals with personality quirks, but they never present any threat to any human character, or, indeed any critique of the status quo of mid-twentieth-century suburbia.[58] With the possible exception of Rosie, they don't have much of any effect on the human characters or their relationships with each other. We may never have flying cars, but the Jetsons' consumerist suburban vision of talking gadgets primed the US public's expectations for chatty household devices. As voice synthesis continued to develop and reach wide consumer deployment in the coming decades, the vision of the future presented in *The Jetsons* might be retrospectively considered to have been unintentionally the most prescient of any pop culture text of the era, although the personality and companionability of the talking devices in the Jetsons' world far exceed those of the voice-enabled technologies of the early twenty-first century. The conversational quirks that Siri may have programmed into it provide only an illusion of personality. We project the rest.

The degree to which the viewing public was invested in negotiating these human/machine boundaries is evidenced by the fact that science fiction was not the only genre to depict computers as talking. As we've established, computer automation, as well as increasing dependence on computer systems for everyday services and activities, brought them to the fore of many people's thinking. One film released at the very end of the 1960s brings together many of the tropes about computers that domestic comedy had developed throughout the 1950s and 1960s: Disney's *The Computer Wore Tennis Shoes*.[59] It is admittedly a less compelling and certainly less artistic film than one like Kubrick's *2001*, but it was popular upon its release, fitting within the slapstick turn in Disney capers like the *Herbie the Love Bug* franchise.[60] A review in *Variety* called the film "amusing" and "above average family entertainment," with "zesty" direction and starring an "adroitly selected group of superior character actors."[61] Unlike stories in which a computer takes on human characteristics, as the title of this film suggests, the main human character takes on characteristics of a computer. This inversion allows the film to frame its central question not as "What is the role of an intelligent machine in human society?" but rather as "What happens when people act too much like computers?"

The plot centers around a Medfield College student named Dexter Riley (played by a young Kurt Russell in his third Disney feature) and his friends, who convince local businessman A. J. Arno (Cesar Romero) to donate his old computer to the college so the students can learn programming, a setup that acknowledges the degree to which computer-programming skills, rather than computer-operating experience, were becoming recognized as potentially integral to an individual's future economic success. In the next scene, Arno opens a panel in the wall to expose a secret back room where, in a robotic male voice, the computer relays statistics from Arno's various illegal gambling pursuits. Whether the screenwriters had Am-Quote or VAB in mind or not, this computer's vocalizations mostly repeat strings of numbers and seem realistically limited in the same way that those synthesizers were, even though the movie's computer is voiced by an actor, not a computer. Arno is quickly established as the villain, as he is running illegal activities in his hidden lair, using the computer's powers of calculation to control what would otherwise be games of chance. Gambling and betting were becoming associated with computers and computational cheating, in another developing television trope.

Arno donates the computer, which is soon being installed at Medfield. The realistic components consist of large metal modules with rows of various colored lights, several knobs, and a couple of monitor screens. Students carry in smaller components and wiring as well, while their professor calls out the numbered parts from a thick instruction manual, trying to get everything in order. This powered-down and deconstructed version of the computer emphasizes its complexity, but also its ability to be mastered by anyone with the right instruction manuals. During his initial lecture in front of the reinstalled machinery, the professor uses simplified, colored charts to teach students the "elements of a computer"—input, memory, control, arithmetic, and logic—with logic illustrated as a cartoon of Rodin's *The Thinker* inside a large red heart, suggesting a combination of intellect and emotion. Although

The Thinker circumscribed with a valentine heart is a comedic touch, the charts actually invoke the cartoon graphics of corporate films about computers from Remington Rand, Bell Labs, IBM, and others that were widely distributed through schools, social clubs, and even theaters in the 1950s and 1960s.

The professor's next chart shows the computer's "output," represented by the picture of a bald head containing a brain with variously colored sections labeled as computer systems, program language, equipment capability, computer instructions, problem analysis, data communications, and operating systems, making visual the metaphor between human brains and computer brains. "Man has done a rather admirable job of imitating the human brain," the professor tells the class. Invoking the oft-repeated highest virtue of Information Age automation, the professor declares that the machine can operate "more efficiently" than people themselves. However, when he attempts a demonstration of how humans can be replaced by computers, the system literally gets its wires crossed.

The experiment itself is not about showing the computer's calculating abilities, but rather, in true cybernetic fashion, its ability to respond to sensory input. When an electronic sensor is triggered by water (meant to represent rain), the computer is supposed to trigger a "program" from its "memory" about what to do when it rains, which in turn triggers a circuit that closes some windows and opens a front door to let a cat in, all of which have been set up in the classroom for the demonstration. The computer then automatically dials a telephone and plays a recording when the phone is answered that orders some groceries to be delivered. The students are impressed and clap enthusiastically at the demonstration. To further the plot, the professor next explains the lineage of the computer and that the computer's memory, shown as large red reels of magnetic tape, contains all the data that it has ever been given. When the professor tries next to demonstrate something with the computer, it goes a bit haywire, banging the window and door from the previous demonstration open and closed until sparks fly out from one of the components and the entire machine shorts out. The students are even more enthusiastic in their applause for this mishap, much to the professor's chagrin. A computer "on the blink" is, by 1969, a comic trope that subtly undermines the authority of computer automation.

The inciting incident of the movie occurs when Dexter returns to the darkened classroom later that night, having run an errand for the professor to buy a part to fix the computer, and, soaking wet from a thunderstorm outside, receives a significant electric shock when he tries to plug two ends of the computer's main power cable together. Dexter holds one end of the cable in each of his hands, his body "completing" the circuit as the computer's lights blink, its control panel beeps, and its tape reels spin.

After a minute, the computer is blown and Dexter is freed from the shock. He runs out of the building, and in the next scene, he is shown asleep in his bed, talking in his sleep *like the computer.* "Three and eight dash zero two dash three zero." He is even beeping like the computer! His confused roommate rouses Dexter awake, telling him matter-of-factly, "You were beeping." A half-asleep Dexter rolls over, unaware. In a comic-book-style inversion of the computer-as-brain metaphor, Dexter has become the computer, robotic voice, electronic beeps, and all. Just as voice synthesis sought to imitate the human voice, this bit of comedy includes the absurdity of a human beeping like the feedback notification of a piece of electronics, a sound that the human body can't make. (In the sound design of the movie, Dexter's "beep" is a sound effect, not the onomatopoeia of a human saying "beep.")

The next morning, Dexter, a middling student at best, breezes through a timed test faster than it would take a person to even read the questions. The scene's sound design makes the most of the human-as-computer inversion. As Dexter's eyes scan the pages, the faint beeping of a computer's "thought process" is heard. When the image focuses on his hand, holding a pencil and checking boxes on the page, the film is even slightly sped up, emphasizing the inhuman time scale of Dexter's progress through the test. As he flips page after page in the test booklet, everyone in the room notices that something is amiss. We eventually find out that Dexter finished in four and a half minutes a test that no previous student has ever completed even in the ninety-minute time allotted. This hyperbole amplifies what was already a ridiculous scenario, but it also sets up Dexter as a parody of the computer's reputation for being able to replace human beings by virtue of their speed.

The professor takes Dexter to a doctor to be examined, and when the doctor takes an x-ray of Dexter's head, the monitor invokes a reverse of the human head chart the professor had used previously to explain the computer. In Dexter's medical image, the shape of the brain is superimposed with layered, close-up moving images of the computer's spinning tape reel, blinking lights, oscilloscope display, and control panel. These are soon replaced by images representing the computer's data in "memory" from its days in Arno's employ—a spinning roulette wheel, bikini-clad models and feathered showgirls, and slot machines. A headline in the next day's paper announces, "Human Computer to Appear on TV," above a picture of Dexter having his head x-rayed. This is a headline that would seem familiar to the movie's adult audience as akin to headlines common during the heyday of the 1950s television quiz shows, when the most successful contestants were sometimes referred to as "Univacs" or described as having computerlike memories. In the film, Dexter is about to be enlisted by the dean for the college's quiz bowl team, but first Dexter becomes an overnight, worldwide celebrity for his "computer brain." He even gets a ticker-tape parade through Manhattan in his honor. Before long, Dexter is also shown being arrogant, selfish, and rude, lacking what today's jargon might call "emotional intelligence." His friend Annie tells Pete, Dexter's roommate, "I liked him better the way he was." The more computerlike Dexter becomes, the less competently he handles his human relationships. Dexter's own "heart" has been replaced by *The Thinker*, and his friends are sad about it.

About halfway through the movie, the main conflict emerges when Arno recruits Dexter to predict horse races for him. This is, again, consistent with one of the key messages, and social conflicts, about computing throughout the 1950s and 1960s. The probabilities that result from statistical processing became "prediction" in the popular vernacular, with all the anxieties that knowing the future raises for human societies. In addition to the marketing from Rand that UNIVAC could help people by predicting the weather and national elections, computers were employed to run data analyses on American football, baseball, and, yes, horse racing. Human anxieties about predicting the future played out in both real and fictional computer matchmaking on television, as discussed in chapter 2. In *The Computer Wore Tennis Shoes*, Arno's racketeering is a symbol of computer-based prediction run amok, while Dexter himself, as the "computer" and a partner in Arno's scheme, needs to be redeemed by the honesty and human loyalty of his friends, whom he has mistreated.

Another character who is taking advantage of Dexter-as-computer is Medfield's dean, who has him competing in a Knowledge Bowl that could mean a large cash prize for the college. When he agreed to compete, Dexter insisted that his three best friends be the other team members, even though none of them are good students. Dexter-as-computer knows all the answers, so it doesn't matter that his friends are useless Knowledge Bowl competitors. Partway through the Knowledge Bowl semifinal, Dexter's answer to a question is the word "applejack," which coincidentally was the password for A. J. "Apple Jack" Arno's file of gambling revenue. When Dexter says the word, he starts speaking in the robotic voice, rattling off Arno's illegal proceeds as the computer had done at the beginning of the movie, complete with electronic beeps. Arno sees this happen and realizes that he has a problem and needs to "put that kid on ice."

Worried when Dexter goes missing (Arno has kidnapped him), Pete and Annie figure out that Dexter is running the computer's memory and that Arno is a crook. In spite of the fact that Dexter hasn't been a very good friend since becoming a computer, Pete and Annie and the gang remain loyal to him. They devise a plan to rescue Dexter and get him to the finals of the Knowledge Bowl, which involves the obligatory silly car chase, during which Dexter is bounced around inside a trunk on the back of a flatbed truck. When his friends help him out of the trunk, his ears are ringing, letting the audience know that both the human Dexter and Dexter's computer-brain are affected by the physical trauma. During the Knowledge Bowl, Dexter struggles to pull answers out of his memory. Like HAL having his memory cores unplugged, Dexter's voice slows and he physically fights to pronounce every syllable, finally collapsing in the middle of answering the penultimate question. With Dexter out of commission for the last question, it seems likely that Medfield will lose, but Dexter's most dimwitted friend and teammate happens to know the answer to the final question, the only answer he's known during the entire Knowledge Bowl—a piece of geographical trivia that he remembers because it involves a place where members of his family live. Human memory may be "slow," but people's tendency to form memories around emotional experiences, rather than memorizing abstract information, is the ability that saves the day. Dexter's "death" as a computer is a rebirth of his existence as a human being, newly aware that his friends' feelings for him make his life as a "regular" human more satisfying than his life as a celebrity computer. (And the bad guys are carted off by the police, of course.)

It may be frothy family fare, but *The Computer Wore Tennis Shoes* uses real concerns about people's social and economic roles in the Information Age to further a ridiculous scenario that nevertheless helps negotiate the boundaries between people and so-called intelligent machines. As much as people were told that they couldn't compete against the efficiency of computers that were gaining new intelligence every day, some people resisted the implication of human obsolescence by asserting the importance of human values and emotions that computers don't share.

DOMESTICATING THE COLD WAR COMPUTER

There is no doubt that the backdrop of the Cold War influenced the way that many people thought about computers in the 1960s; one of the computer's main raisons d'étre was creating and controlling weapons of mass destruction in World War II, and their use in civil defense throughout the Cold War caused as much paranoia as feelings of safety. In his insightful analysis of Cold War computing, Paul N. Edwards explains American Cold War ideology as the circulation of "closed-world discourse," or articulations between language, technologies, and practices that together "supported the visions of centrally controlled, automated global power at the heart of American Cold War politics."[62] Edwards describes *2001*'s *Discovery One* spaceship as an example of this discourse, as the quintessential "closed world" environment where information-processing computers constituted a "dome of global technological oversight . . . within which every event was interpreted as part of a titanic struggle between superpowers," and where HAL 9000's voice represents "the faceless, nonlocalized, uncaring power of ubiquitous high technology."[63] It is certainly a film that asks provocative questions about key ideological themes of its time (if not our own).[64]

However, sometimes in their enthusiasm for the reverence that Stanley Kubrick seemed to have held for computing technologies, many engineers, journalists, and even cultural critics who have written about talking computers have often assumed that Kubrick's vision of a single-minded, autonomous AI was more dominant in its cultural influence than it actually was.[65] In *Star Trek*, that other favorite science fiction text of the 1960s, ideas about both what computers could do and what they should do played out very differently, as the *Enterprise* computer was completely under the control of the starship's crew. In family entertainment like *The Computer Wore Tennis Shoes*, the limitations of computers and the abilities of human beings also hinged, as in *Star Trek*, on computers' unemotional nature, on the incommensurate differences between electronics and human embodiment, and on the unsuitability of applying computation to relational issues between people.

As with HAL 9000, these alternative versions of computing power also used vocalizations from computers to audibly represent these negotiations. HAL's omniscience and fully conscious AI is the most fluent, speaking at will and without sonic electronic artifacts, maintaining a calm monotone representative of "his" efficiency and logic. The *Enterprise* computer, in contrast, is just called "computer," only speaks when spoken to, and only for the purposes of recording data or providing an answer to a request for information. It is the tool imagined in Eames's *The Information Machine*, but with a robotic female voice that represents the unsexed subservience of pink-collar information work.[66] In one episode, the computer has received an upgrade and starts addressing Captain Kirk as "dear" in a sultry voice, an issue that Kirk regards as "a serious malfunction."[67] The character of Dexter inverts the question of computer/human relations by turning a human into a computer and giving him the absurd vocalization of electronic beeping to represent the radical incommensurability of machine and human embodiments. While its simple "good guys versus bad guys" plotting and silly play on computer and human "memory" are certainly less sophisticated than Kubrick's metaphysical musings, in some respects that absurdity is exactly the point. In navigating human relationships in day-to-day life, talking computers only get in the way.

4 SPEAK & SPELL (1978)

The most famous voice synthesizer of the twentieth century wasn't used to make a computer talk. Instead, it was the voice of an educational game for children: Speak & Spell, from Texas Instruments (TI). Children in the late 1970s heard Speak & Spell say, "That is correct," in its tinny glottal rumble, over and over again on television commercials aired during their Saturday morning cartoons. Speak & Spell's tagline, "Learning should be fun!" fit right in with *Sesame Street* and *Schoolhouse Rock!*

Speak & Spell is now an icon of Gen-X childhood, with derivatives of the toy popping up all over in popular culture: helping *E. T. the Extra-Terrestrial* phone home (1982), as Mr. Spell in Pixar's *Toy Story* (1995) and the playmate of the possessed doll Chucky in *Bride of Chucky* (1998), and helping the *Penguins of Madagascar* to order takeout (2008), just to name a few. Speak & Spell's distinctive voice has been included in numerous pop music recordings across many genres, including cuts by Orchestral Manoeuvres in the Dark, TLC, Coldplay, Limp Bizkit, Beck, Röyksopp, and the quintessential technoband Kraftwerk, who featured it in the title track of their *Computer World* (1981) album. In that track, Speak & Spell's voice repeats the words "business," "numbers," "money," and "people," the sterile implications of which are juxtaposed against another synthesized voice at the end of the album that repeats, "It's more fun to compute."

In the late 1970s, Speak & Spell was part of a large wave of microprocessor-based consumer electronics that made interacting with digital devices part of everyday life, especially for children and teens. Spurred by the popularity of blockbuster movies including *Star Wars* (1977), *Battlestar Galactica* (1978), and *Star Trek: The Motion Picture* (1979), the market for electronic toys and games amounted to about $150 million in the US in 1978 but ballooned to more than $500 million in 1979.[1] Most of these incorporated beeps and bleeps like R2D2, if not a robotic voice reminiscent of C3PO, conditioning young users to expect the kinds of voice interactions from their electronics that they saw on the big screen. Some parents were concerned about the amount of time their children spent gaming (at least those who weren't using electronic chess, backgammon, and sports games of their own), but they also recognized that their children's economic futures were going to depend on the ability to use computers. Boosters promoted Speak & Spell and other programmed electronics as excellent preparation for this future, presaging the adoption of personal computers into middle-class US homes in the next decade.

As an object designed for school-aged children, Speak & Spell followed trends for positioning computers as educational, not only as business machines.[2] More specifically, its voice synthesis feature was an industry-wide proof of concept for what the interface for home computers could become for the entire family. Voice-as-interface might expand who could use a computer, making it potentially as easy as talking on the phone. It could also make using lots of other electronic products easier, from copy machines to cars. In spite of the fact that TI's voice synthesis fell far below the C3PO benchmark for both understandability and personality, it was imagined as the voice of the machines of daily life for a brief time, and proof that machines could have voices for years to come.

This chapter follows the development of the Speak & Spell, the first widely available consumer electronics device to feature digital voice synthesis, to show how market demands affected the engineering and promotion of voice synthesis technologies at this time. Just ahead of software-based synthesizers that will be discussed in later chapters, TI's speech synthesizer-on-a-chip and its imitators put dollar signs in the eyes of some businesspeople and analysts but actually underwhelmed consumers. Instead of talking household and business gadgetry, the real take-up of voice synthesis was in digital games and toys, having a significant impact on children's expectations for the world they were growing up in. The development of the Speak & Spell at TI follows the digital signal processing (DSP) achievements discussed in chapter 3, but it shows how the constraints of designing voice synthesis at a consumer price point affected the sound and function of the device. Nevertheless, the wider consumer context in which the Speak & Spell was introduced, including the rapid rise and fall of handheld digital games in the market, has had a lasting cultural impact, especially on the generation of young people who were the first to spend time in front of digital screens as leisure activities and at school. The case study of Speak & Spell provides evidence that personal computers were poised to take up roles as *educational* machines, not only business ones. Like the teachers that they were going to supplement (or replace, in some people's minds), these machines were going to talk.

INVENTING THE SPEAK & SPELL

The first Consumer Electronics Show (CES) was held in New York City in June 1967. As a spin-off of the primary trade show for consumer audio equipment, it's no surprise that of the 117 exhibitors, many debuted turntables, transistor radios, and players for the new compact audiocassettes. The show was immediately successful, with more than 17,500 attendees in 1967, and moved to a semiannual schedule in 1973. Color television in the late 1960s, as well as the debut of the first videocassette recorder in 1970, the Laserdisc player in 1974, and the VHS player/recorder in 1975, helped make CES the destination for new televisual gadgets, as well as consumer audio. The early 1970s development of microprocessors, wafers made of silicon that contain all the functions of a computer's central processing unit on a single integrated circuit (IC) or a small number of ICs, expanded the diversity of electronic gadgetry even further, with digital pocket calculators and video games filling booths at CES, and then store shelves during the 1970s.

One of the most commercially successful microprocessors of the time was the 4-bit TMS 1000, introduced by TI in 1974 at a price of less than $3 each in bulk quantities.[3] In addition to its chip business, the success of TI's first commercial calculator, the TI-2500, introduced in 1972, prompted the company to add to the offerings of its own consumer products division. A reverse calculator for children, the Little Professor, went on sale in 1976 for around $20 and proved very popular, and profitable. Attempting to replicate that success, a few TIers met to brainstorm new product ideas along these same lines, and the concept of a spelling aid was tossed out. Paul Breedlove, a project manager recently transferred from Scientific Calculators to Consumer Calculators, was tasked with investigating the feasibility of the idea and outlined the design for a handheld, talking "spelling bee" in his design notebook.[4] Breedlove had previously worked in TI's Speech Research area and assumed that a talking toy wouldn't be much of a stretch of the available voice synthesis technology.[5] In fact, designing a complete voice synthesizer for a single IC was improbable enough that TI's corporate office turned down Breedlove's request for "Wild Hare" funding to develop the idea because it was "too wild."[6]

For the purposes of a learning aid for school-aged children, the single-word paradigm of previous voice synthesis applications made the possibilities for using voice synthesis more feasible. Still, as one of the Speak & Spell developers would later recall, "At the time, the state of the art for implementing real time speech synthesis in a single integrated circuit (IC) was well defined: it was impossible."[7] A four-man team persevered and were able to finally convince TI to fund the project after demonstrating a proof of concept using a minicomputer. The talking Speak & Spell, which had shed its original bee shape for a cheaper-to-manufacture boxy design, was unveiled at the Summer CES in June, 1978.

The next day, the *Wall Street Journal* covered "Electronic Gadgets That Can Talk," quoting a Morgan Stanley analyst that voice synthesis heralded a major new direction in consumer goods.[8] Meanwhile, an energetic public relations rep for TI put Speak & Spell on a press tour, getting Jane Pauley and Tom Brokaw to spend three minutes highlighting it on NBC's *Today* show, with Brokaw holding up a cutaway model in order to "play up the idea that the Speak & Spell [did] not use a tape-recorded message, but is something like a computer."[9] Only three months later, a picture of Speak & Spell took up the entire cover of *Business Week* magazine for September 18, 1978, along with the headline "Texas Instruments Shows U.S. Business how to Survive in the 1980s." Voice synthesis had officially moved from the science news to the business pages.

Speak & Spell was the first widely available consumer application to benefit from the digital voice synthesis processes worked out in research labs during the 1960s as described in chapter 3, while its microprocessor was a product of the achievements in electronics miniaturization driven, in part, by the space race. By the end of the 1950s, transistors were ubiquitous, but there was a limit to how small they could be. Research on several fronts pursued the idea of building circuits as single devices containing all the necessary parts, in miniature. Bell Labs had made the leap from vacuum tube to transistor, but it was Jack Kilby at TI and Robert Noyce at Fairchild Semiconductor who brought the IC to fruition.[10] Electronics, shorthand for gadgets that contain circuitry for controlling electrical energy, weren't born with the IC, but they became more affordable, more powerful, more diverse, and more ubiquitous because of it.

Speech wasn't a gimmick for the Speak & Spell engineers. When Breedlove first imagined an electronic spelling bee, he recognized that "for the same reasons children search out a cooperative friend or parent to call out spelling words, a voice would be needed for the product."[11] The product needed to give the user a word without giving away the spelling of the word, and the most obvious way to accomplish this was to design a speaking machine. The only problem was that the state of the art for voice synthesis at this time required more computing memory and processing speed than could be achieved at a consumer price point. In the September 1972 issue of *Scientific American,* the inventor Richard T. Gagnon had advertised Votrax, claiming it was the "world's first low-cost, miniaturized voice synthesizer" at "a cost of under $2,000 in production quantities," though, even in its third generation a few years later, the Votrax voice was barely understandable.[12] One journalist described it as "just adequate for a motivated person to be able to understand what was said."[13] Breedlove's team had their work cut out for them.

Richard Wiggins, the member of the Speak & Spell team responsible for the speech synthesis, decided on linear predictive coding (LPC) as the synthesis method because of its computational efficiency.[14] In 1978, the standard memory for a consumer device was only about 4K. *Popular Science* reported of the Speak & Spell in 1980, "The device has stored in it 200 words, about 3½ minutes of speech. Not very impressive—until you realize that the 256,000 bits of digitized data are all compressed onto a pair of integrated circuit chips the size of a thumbtack," making much of the technical feat.[15] The digital processes were still described by analogy to the human body, as previous voice synthesis had been. Following Atal and other Bell Labs researchers who first published the explanation of LPC, Wiggins also explained: "The process has certain similarities to the actual mechanisms of human speech, in which the vocal cords have air forced past them by the lungs and the rest of the vocal tract (tongue, teeth, lips, etc.) modifies the resulting sound."[16] Among several that were filed, perhaps the most valuable of the patents to emerge from the Speak & Spell project was number 4,209,836, awarded to Richard Wiggins and George Brantingham for a "Speech Synthesis Integrated Circuit Device." Tables that include the entire instruction set swell the patent to over 150 pages, but the claims are only 11, and describe a digital filter speech synthesis circuit responsive to "a plurality of digital values" representing voiced and unvoiced speech, pitch, amplitude, filter coefficients, digital-to-analog conversion calculations, timing, and signals from the keyboard input.[17] The digital vocal tract model hadn't changed much over that of the electronic vocal tracts (EVTs).

The Speak & Spell followed the simple electrical model of voice production worked out at Bell Labs: a source of sound consisting of both a noise source and a pulse source to enable voiced and unvoiced sounds, a filter modeled from measurements of the human vocal tract, and the voice output at a speaker "mouth." TI's patents make no mention of these body analogs, but their lineage in the prior art is evident. In other documentation, the Speak & Spell's synthesizer is described alongside a block diagram that identifies the sound generators as part of the "vocal cord model [sic]" and the filters as part of the "vocal tract model."[18] An image of a human head is provided for reference. In describing their specific model, the engineer Gene Frantz explained: "The reasoning behind the variation in bits per reflection coefficients has to do with the part of the human vocal tract each coefficient represents. K1 can be thought of as representing the lips, K2 the teeth, and so on where K10 represents the back of the throat. Obviously, the teeth and lips have more movement than the back of the throat and therefore have more bits assigned to them."[19]

The assumptions hearken back to Dudley's model of fixed or variable parts. In fact, most of the TI team's assumptions came straight out of Bell Labs research. In discussing one of the decisions they made, Frantz stated, "We chose 8 [kHz] as the right sample rate as that is what the phone system used. The issue is that many of the unvoiced sounds are actually above 4 [kHz] (the bandwidth of the signal is, by definition, less than half of the sample rate according to the Nyquist theorem)."[20] The innovation of Speak & Spell was to get voice synthesis on a single, affordable chip. It was basically a tiny, programmable signal processor.

Technical limitations influenced the design of Speak & Spell at every level. In addition to the 8-kHz sample rate, which offered a compromise between data rate and voice quality, Frantz listed the frame rate, multiplier size and data word size, number of coefficients, and bits assigned to each coefficient as decisions that had to be made to balance the understandability of the voice with the memory and processing speed available. Frantz even explained that the design used "tricks" to optimize what was available. He reflected:

> The first time I told Richard that I was not going to use an anti-aliasing filter in the design, he got a bit irritable. . . . I had decided that the anti-aliasing filter came free with a cheap speaker. The speaker I chose was a $0.50 two inch speaker that has a frequency bandwidth of 300 Hz to 3.3 [kHz] . . . . In my discussion with the speaker vendor, I told them that if they ever sent me a speaker with a better frequency response than the specification I would send it back. I truly needed all of the "features" of a cheap speaker to make the system cost effective.[21]

The "greatest technical hurdle," according to Frantz, was finding an "LPC friendly" voice to record the samples.[22] Given the sample and data rates in their LPC implementation, the higher formants of a female voice could be missed, so it was determined that a lower-pitched voice would work better. Although the choice of gender was entirely technical, Frantz recalled explaining the use of a male voice in sexist terms "over the years" whenever he presented this aspect of the Speak & Spell development: "Because the pitch of a woman's voice is higher than a man's voice, the spectral lines are further apart in the frequency domain. Another way of looking at it is that there isn't as much information in a woman's voice as there is in a man's voice, given a constrained bandwidth. Perhaps that is why women talk more than men—trying to make up for the lack of information in their voices."[23]

"Although it doesn't accurately explain the reason we couldn't do female voices, the people in the audience always remembered the explanation," justified Frantz. Another point that it takes advantage of, though, is the conflation of the technical and colloquial definitions of information. In fact, just as with the Am-Quote and Voice Answer Back (VAB) systems of more than a decade before, the voice needed to be low and monotonic. The team found a voice actor who could do great character voices, but the character voices didn't fare any better when synthesized than higher-pitched voices. A quick marketing survey determined that children didn't have much of a gender preference, so the voice samples were eventually recorded by a male Dallas-area DJ in a low, vibratoless monotone. Even so, the DJ's west Texas accent occasionally had to be edited frame by frame to accommodate the system's constraints, as well as to fit the parameters of what the team had determined would be the official voice of the Speak & Spell: the standard American broadcast accent as represented by the pronunciations specified in the conservative *American Heritage Dictionary* (1969). Therefore, Speak & Spell ended up with a highly standardized malelike voice, with electronic artifacts. Further compounding the standardization was the fact that Speak & Spell had only one volume. A volume control knob would have added a dime to the production cost, a concession that design engineers chose not to make.

Both synthesis and memory constraints also affected the words that could be included in the spelling lists. Speak & Spell used read-only memory (ROM), some of the largest available at the time with 131,072 bits, but this was only enough to store about 150 words.[24] That problem was solved by selling additional ROM cartridges that included expanded word sets, which could be swapped into the machine through the back. However, there were additional constraints on the words that could be used. "Intelligibility" was important because words would be given without any context (no accompanying image, no "Can you use that in a sentence, please?" feature). You only got the machine's pronunciation of the word as a clue. For simplicity, the initial word lists focused on single-syllable words. Some phoneme combinations just didn't work, especially unvoiced consonants, which have higher frequencies. Frantz explained a couple of the problem words they ran into: "Specifically the unvoiced sounds f, s, t, p, h, sh, were sometimes misinterpreted as one of the others. Here are some examples: The word four was easily misunderstood as 'hore' if not spoken in context. We needed to keep the number four but chose not to use any of its homonyms in the vocabulary list (e.g., for, fore). In a later product we realized that [the letters] p and t could be misunderstood causing a specific problem with the word ship. We ended up canceling the product because of this issue."[25]

As engineering often goes, this bug resulted in an added feature: one of the games on the final product was created out of concern for the intelligibility issue—the "Say it" mode. The engineers concluded that it would be wise to have a mode where the child could hear the word while looking at the spelling in hopes that when the word was spoken in the "Spell it" part of the game, it could be easily correlated to one of the spelling words seen in the "Say it" part of the game. According to Frantz, educators really liked this partnering of audiovisual training of the spelling of words, but it was a feature that emerged as a solution to a technical constraint rather than an educational imperative, and one that potentially disciplined child users to the vocal limitations of the machine.

These small details about the decisions that the engineers had to make demonstrate that the technical constraints of brute force engineering accounted for many of the characteristics of the device and its voice. The engineers included parents and educational consultants in the design process, but almost entirely ignored their opinions. The spelling expert and textbook author William Kottmeyer was retained as the educational consultant for Speak & Spell. After his first meeting with the engineering team, Kottmeyer recommended killing the product. His main objections were the capital letter display, the alphabetical layout of the keyboard, technical limitations on the vocabulary, and the reinforcement of a rote method of learning.[26] As Frantz recalled, "We listened to him very carefully and then ignored his ultimate recommendation."[27] In some cases, the consultant's advice "obviously . . . had to be ignored" because the engineers were already focused on the technical breakthroughs that they were making.[28] In other cases, the engineers justified their instinct, as in the "rote learning" aspect of the device. They argued that the innovation for education was, precisely, taking a conventional activity, one with a clear "algorithm"—adult says word, child spells word, adult gives feedback about correctness of child's spelling—and putting it in an electronic learning aid (i.e., automating it).[29] In fact, the Breedlove and Moore patent for "Electronic Learning Aid or Game Having Synthesized Speech" cites 1950s-era behaviorist teaching machines as prior art, an important point to keep in mind when the product is later lauded by some educational experts as being "engaging" in a way the behaviorist model was supposedly not.[30]

In addition to behaviorism, midcentury assumptions about universal patterns of communication are also embedded in Speak & Spell's design. The idea of creating an automated algorithm to replace what was previously a face-to-face interaction separates language from communication, deemphasizing the value of a person's embodied expressiveness by turning voice into an instrumental, standardized medium for conveying words. It also shows the influence of Noam Chomsky's universal grammar theory of linguistics, which separates syntax and semantics as information theory does, and asserts that linguistic forms are based in the brain and thus lead human beings to develop language with standard properties. Universal grammar provides a theoretical basis for establishing communication as computational and was foundational to both computer science and cognitive science. It continues to serve as a language model within these fields in spite of its significant limitations.[31] Universal grammar supersedes any notion of human communication as negotiated and dynamic, culturally specific, and embodied beyond a computational model of mind. The spelling practice scenario wasn't understood as an act of human communication, a relational experience between people; it was just an algorithm that could be automated by a machine.

Speak & Spell also provides insight into how social scripts are written into the speech acts of talking machines. Because memory and processing power were premium resources, engineers needed to keep things simple, but they also had specific social motivations for programming what the machine's responses would be. Speak & Spell didn't use any words to refer to itself, but its illusion of agency could be implied from the way that it commanded the user to "say it" or to "spell it," by the way that it judged the user's input as "correct" or "wrong," and by the way that it served as a proxy for the parent or teacher who might have previously quizzed the child on their spelling. The engineers thought of the linguistic situation as purely functional, even though they marketed the device as "fun." Indeed, as has been described, they had little choice in the matter, as understandable sentence-length speech was difficult to achieve sonically and cost much in terms of memory and processing. So Speak & Spell was designed as a strict teacher. On a first misspelling, the child would hear a direct and unemotional, "Wrong. Try again." If the child misspelled the word on the second try, Speak & Spell responded with, "That is incorrect. The correct spelling of . . . is. . . ." While the engineers considered other kinds of responses, including raspberries or humorous comments, they inevitably decided that Speak & Spell should not "reward" incorrect answers with entertaining responses. Frantz called the "personality" of Speak & Spell "engaging," but really, Speak & Spell had nothing that could be called "personality" at all, just a monotone voice and a small set of short, programmed responses. "That is incorrect" was likely to get frustratingly repetitive for a young child, even though it was efficient and functional.

The Speak & Spell engineers discounted the opinions of focus groups in the same way, and for the same reasons, that they had ignored the educational consultant. Before a proof-of-concept model was available, a couple dozen mothers with target-age children (seven to ten years old) were presented with a conceptual description of what Speak & Spell would do, cobbled together from tape recordings of how the product might sound, sample words, and a talking pull-string toy as an example. The moms were no more enthusiastic than the educational consultant had been, worrying that the product would be boring, noisy, unreliable, and, most of all, expensive. When offered seven possible character voices for Speak & Spell, the women liked a "computer" voice best, saying that it was silly and believing it most likely to engage their children, (although another option, called "Andrew" in the notes, was also agreeable if it could be made understandable). None of the other characters gained any real traction with the moms. At least one person was concerned that none of the voices was clear enough to be understandable to a child. In one focus group of nine women, seven preferred the talking computer option.[32] However, the engineers "didn't want to have to explain to the technical community the reasoning for having created such bad speech synthesis,"[33] so they worked to give Speak & Spell as "real" (i.e., "regular" human) a voice as possible. It's notable that the moms preferred the talking computer option, as that suggests the degree to which talking computers were already a cultural touchstone even for children.[34]

Frantz believed that the focus group participants "missed the concept" altogether, reasoning that Speak & Spell "solve[d] a problem that no one even knows they have. . . . The leap . . . was too far for them to grasp."[35] That's not entirely true. It seemed that the women in the focus groups understood perfectly what the device would do—they just didn't think that it would hold children's interest for very long, and they weren't willing to bet the $70 price tag on that possibility. Frantz thought that perhaps the parents' limited experience with talking machinery, based as it was on "movies where the talking machine was cast as a bad guy," also tempered their enthusiasm for the product and "reinforced the existing ignorances and prejudices of the consumer concerning talking toys."[36] However, as we've already seen, there were a wide range of talking machines in movies and television up to 1978, and only a few of them (and they were least likely to be seen by children) presented the machine as a "bad guy." Besides, there is nothing to support that interpretation within the focus group notes. There is no mention of HAL or any other fictional computer. At least one person mentioned being disappointed with the quality of previous talking toys, though given that the hottest selling toys at the time were mostly electronic, there's little to support widespread "existing prejudices concerning talking toys." One comment even suggested that Speak & Spell would be worth more if it included cassettes that could contain more than 250 words, although others suggested that tapes wore out too quickly.

It wasn't the speaking aspect of the device that most parents were hesitant about; it was what little value you seemed to get for the price. (Parents recommended a price tag of no more than $25. Even when discounted for its second Christmas on the market, Speak & Spell cost twice that.) In fact, their suspicions were probably well founded. There were very few studies about the educational value of Speak & Spell once it was available, even though some school districts provided them to classroom teachers. In one notable exception, British educators C. D. Terrell and O. Linyard found that students given a Speak & Spell for two weeks showed an initial improvement in spelling performance on words in the machine's lexicon. They also found that this improvement was transitory, and once Speak & Spell was taken away, students' performance returned to previous levels. Also, they found no statistically significant improvement in the spelling of words not in the machine's lexicon.[37]

One thing that the focus group discussion did surface was parents' anxieties that their children needed to "understand computers." Even ahead of the personal computer revolution, parents were aware that their children's economic futures were going to be closely tied to computing devices and the ability to effectively use them. "Children don't think of computers like we do," one woman is recorded to have said. "Children need to learn computers," stated another. As a machine for children, Speak & Spell might have paved the way for the home computer to be promoted as accessible for all ages, as well as a need of the entire family, not only for home management or a parent's business. It might also be seen as an initial step in changing the dynamics of postwar play from imaginative to more structured and goal focused.[38] It was intended to automate homework—offloading the participation of parents onto a machine and creating a new market category for home-based educational technology.[39]

The historian Carroll Pursell observed that all technology can be played with, but he reminded us that play and technology are historical and contingent categories.[40] Pursell found that, with regard to toys, children are encouraged to follow the scientific and technological fads of society. For example, ecology kits were marketed during the environmental movement in the 1970s, a toy submarine was patented in 1920 and a toy telegraph was available in 1930, and even a Tonka truck assembly line kit was on the market in 1936 for future Fordists (to say nothing of the various iterations of toys that mimic household appliances throughout the generations).[41] At this point in the late 1970s, one of the key arguments for purchasing children's electronics was to prepare children for an increasingly computer-centric future that was very quickly approaching.

PLAYING WITH SPEAK & SPELL

The IC speech synthesis chip that the TI team developed, the TMC0281, garnered many accolades from the technical press. These started right away, when *Industrial Research and Development* magazine honored Speak & Spell as one of the most significant new technologies of the year in 1979. The Institute of Electrical and Electronics Engineers (IEEE) dedicated a milestone plaque at TI's Dallas site in 2009 for the first use of a DSP IC for speech generation. Also in 2009, the *IEEE Spectrum* magazine listed the TMC0281 as one of "25 Microchips That Shook the World."[42] However, there was surprisingly little popular buzz about Speak & Spell when it debuted, in spite of a concerted marketing campaign that featured Bill Cosby, comedian, doctor of education, and *Advertising Age* Spokesman of the Year for 1977. Given its iconic status since, this seems remarkable. For consumers in the late 1970s, however, Speak & Spell was one of an avalanche of handheld electronic gadgets blinking and beeping from store shelves. Speak & Spell's speech synthesis chip may have been a significant milestone for the tech and business communities, but for average consumers, mostly overwhelmed parents dealing with an ongoing fuel crisis and economic recession, its voice was just one of many emanating from the barrage of merchandise that was being peddled to children from the television set. Even though most of these talking toys spoke a handful of words at most, and usually utilized old phonograph technology to do so, the cultural imagination for talking machines was already well developed through watching television and Hollywood movies. Parents, who were inevitably responsible for making purchasing decisions, were products of the cultural imaginary that was formed in the 1950s and 1960s as discussed in the previous chapters, including many pop culture representations of computers and robots, as well as an educational system increasingly focused on what we now call science, technology, engineering, and mathematics (STEM).[43] But their children—the cohort labeled Gen X—would be the first to have electronic toys and games specifically marketed to them (and the first generation to have computers in their family rooms and classrooms). By the time Speak & Spell was announced in 1978, electronics had suddenly taken over the toy and game industry.

Playthings, the industry magazine for toys and hobbies, reported in a short blurb in January 1977 that an upstart sector of the games business—electronics—was predicted to become a "major sales category" due to the "growing use of microprocessors," which was "expected to advance their technological finesse to a greater extent." The news brief rather awkwardly stated that "the stand-up (arcade) coin-operated electronic games, along with video television cocktail table games, are expected to produce the largest future growth," a figure projected to be over $600 million by 1985.[44] The industry was new enough that *Playthings* reported "virtually all home electronic games are variations of Atori's original 'pong' game [sic]," but that they were expected to become more sophisticated.

Browsing through ads of toys available in the 1977 toy fair preview, one saw lots of battery-operated gadgets that mixed familiar characters and new features. These included a Barbie typewriter and a Holly Hobbie sewing machine, both a contemporary push-button telephone set with a working intercom and a Holly Hobbie candlestick "talking telephone" with eight different sayings, and a Lone Ranger shooting arcade tabletop game and a racecar pinball machine, both mechanical but with moving parts and some lights for bling. A battery-operated Robot 2500 could move forward slowly and resembled a miniature Westinghouse Elektro, suggesting that android dreams hadn't changed all that much in the preceding forty years. Many of the offerings in the toy fair preview section were voiced in various ways. In addition to the talking Holly Hobbie phone, one manufacturer was hawking a "see & hear" toy camera, and there were plenty of toy radios, many disguised as something else, like a stuffed animal or a handbag. Cassette recorders, citizens band (CB) radios, and walkie-talkies—even *Sesame Street* branded—were now marketed as toys. And almost everyone was introducing electronic handheld games. Tomy had a pocket-sized game called Blip that was advertised as "almost like Pong,"[45] and Mattel's Auto Race and Football had players controlling a "light blip (representing a car or ball carrier)."[46] Milton Bradley boldly announced, "It computes!" to bring "the marvels of computers and electronics" to a new line of games.[47]

All the major toy manufacturers were introducing electronic toys for all ages. Ideal, the company that got started by importing plush bears named for

Teddy Roosevelt, said that it was mixing "the marvels of modern technology, the pleasures of simple play, and encores of memorable playthings from the past" in introducing its 1977 line, which included Electro Man, a superhero action figure "out of the world of computer electronics."[48] Creative Playthings was flying into the electronics age with Jet Pilot, "an electro-mechanical plaything" that resembled a flight simulator with four available flight plans for the school-aged, and the Little Maestro Piano-Organ with twenty-five keys for the little tykes.[49] Coleco had a freestanding Fonz pinball, handheld football, and a two-player video tank game called Telstar Combat (so named because it was a console that allowed game play on a television set).[50] The consoles proved to be a little too far ahead of the curve, losing $14 million in 1977.[51]

In addition to the miniaturized radio, tape recording, and phonograph technologies for making things talk, inexpensive voice boxes could be incorporated into almost anything. Kinter Products, one of the parts manufacturers providing these low-cost talk boxes, advertised a battery-operated unit that could play eight short messages or one long message with "Hi-Fi sound fidelity," and a smaller pull-string unit that could play six messages in "any type of voice."[52] One of its competitors, Ozen Sound Devices, claimed sixty-one patented devices, including some as small as a human thumb and others operated by gravity change.[53] These were the kinds of devices that gave voice to Mattel's Hug 'n Talk doll, Bugs Bunny See 'N Say Talking Telephone, Talking Vue toy camera (sixteen "shots" and six audio messages), Talking Magic Crystal Ball, and many, many others.

But the Christmas season sales winners for 1977 were anything connected to *Star Wars*.[54] So when Speak & Spell showed up for Christmas 1978, it was accompanied by an onslaught of robots, droids, spaceships, action figures, laser weapons, and aliens, along with a Santa's sleigh full of handheld electronic games. In December 1978, electronic games, or "Turned-on Toys," made the cover of *Newsweek*. Best-sellers, including the handheld Merlin and the tabletop Simon were highlighted, as were two chunky robots designed for preschoolers: "talkative" 2-XL by Mego and Playskool's Alphie.[55] These new "cybernetic wonders" came in all shapes and sizes, but they had in common that they were "run by computers no bigger than a stick of Dentyne gum."[56] Often, this "stick-of-gum" computer was an iteration of TI's TMS1000, the first available high-volume microcontroller that included a 4-bit central processor, ROM, and input/output lines on a single chip. It may have been TI's speech synthesis chip that won all the technical accolades, but it was the TMS1000 that was the bread and butter of TI's IC business. Introduced in 1974 for less than $3 each when purchased in bulk, TI sold more than 25 million of them in 1978. However, shortages were a constant problem, much to the consternation of retailers.

It doesn't follow, though, that average consumers understood the differences between toys and games that could blink and talk and those that contained microprocessors that controlled that blinking and talking. Even within the *Newsweek* article purportedly *about* "electronic toys run by computers," some of the toys are electronic, not computational. With the announcement of the commercial availability of microprocessors, computation took on a new cloak of invisibility, as "computers" were now embedded inside products that most consumers would never tear down. When the *New York Times* ran only its second article to use the word "microprocessor," in January 1975, the technology columnist Victor McElheny described reducing the size of "wires and switches used to transmit electronic information" to a fingernail and quoted the Hewlett-Packard vice president for research Bernard Oliver as explaining, "the central processing unit, the dominant part of the computer that you used to get billed by the second for using, has shrunk to the point of invisibility."[57] As deployed in "automobiles, gasoline pumps, traffic signals, and supermarket cash registers," these invisible computers were expected to "impact daily life," but not in a way the general public would necessarily *see*, even if hobbyists did. In the mid-1970s, "computers" became "chips," but not everything that contained a chip fell within the popular paradigm of "computer." One way that this difference was made "visible" was in the price, with some parents more than happy to purchase a less expensive "robot" that talked via cassette tape over a more expensive one that contained a programmable microprocessor.

For its article, *Newsweek* had gathered a focus group of five- to nine-year-olds and had them rate a few of the season's toys. The popular electronic pattern-matching family game Simon and a 2-XL robot from Mego were the winners, with another remote-controlled robot named Sir Galaxy, which functioned somewhat like a walkie-talkie, also interesting the kids. Although Sir Galaxy's voice was that of the child talking, 2-XL did talk on its own. Mego advertised it as "the toy with a personality" and "the talking robot with a mind of its own." *Newsweek*'s description makes the toy seem deceptively sophisticated: "2-XL robot cannot walk. But the $50 Mego toy does talk—in a pleasant, lively voice all its own. 2-XL asks true-false questions, correcting wrong answers and congratulating the respondent for right ones. It also gives multiple-choice tests, cracks amusing jokes and plays a few games." The kids liked it because it told jokes; "It's almost like a really little person," a boy is quoted as saying.[58] Advertised as "the robot that thinks he's human," 2-XL actually used 8-track tapes for its voice (there were thirty-six to choose from, for all ages, including exercise tapes for mom). Four buttons could move the tape track, making the robot minimally interactive.[59] The kids were less impressed with the microprocessor-controlled Merlin, whose games, including Black-jack, were a little beyond them. Merlin was marketed for ages seven to adult; Parker Brothers called it a "hand-held electronic wizard with [a] powerful computer brain" that had a "vocabulary of space age sounds" and lights that communicated wins, losses, and ties.[60] The company included question-naires with its initial release of the game and claimed that "hundreds" of them were returned with a 95 percent positive response.[61] However, one of *Newsweek*'s kid testers stated of Merlin, "This machine cheats."[62]

Merlin was designed by two former National Aeronautics and Space Administration (NASA) employees, Bob and Holly Doyle, who saw an opportunity to use their computer expertise to design products for the bur-geoning consumer electronics market. This fact is symptomatic of the sig-nificant changes that microcontrollers, and the potential demand for toys that utilized them, meant for the toy industry. There were about two dozen games and toys that incorporated microcomputer chips in late 1978, and they were hard for consumers to find by the holiday-buying season. Retail-ers hadn't anticipated the popularity of this new category when they'd been shopping at February's toy fair. Mattel dealt with the resulting consumer backlash by absorbing the cost of sending extra units to retailers who couldn't keep them in stock. Then 1979 would bring its own problems, as a rise in microcomputer-chip-enabled toys and games (about 115 options by the end of the year) caused shortages from chip manufacturers, including TI.

Electronic games were finally the *Playthings* cover story in April 1979 (although the editor's column focused on model kits, beginning, "In all the hubbub over the electronic category, the trade must not forget the product lines which are the foundation for the day-to-day operation in toy and hobby departments").[63] Over 120 new electronic products were introduced at the 1979 American Toy Fair, "a bonanza . . . staggering to behold," according to the bewildered trade press, which recognized both the potential and "a great deal of unanswered questions about specific problems" faced by every area of the toy trade, including the pricing of products and materials (the oil crisis was also contributing to a shortage in available plastic), marketing, and impacts to traditional categories, with board-game producers particularly concerned (especially since it might cost eight or ten times more to develop an electronic game than a board game).[64] Parker Brothers research and development chief Bill Dohrmann explained, "We'd spent 90 years pasting paper on cardboard, and suddenly we're dealing with computers."[65] Consequently, many of the new IC products were offered by companies that were newcomers to the toy industry, like TI.

Even though TI was very specific about it being a learning aid, it was not uncommon for consumers to see Speak & Spell advertised next to the full spectrum of electronic games available in late 1978: Coleco's Quiz Wiz, Amaze-A-Tron, and Electronic Quarterback; Parker Brothers' Merlin; Mego's Electronic Baseball; and Milton Bradley's Simon and Comp IV. It was also common for Speak & Spell to show up in toy store ads that didn't differentiate between Mattel's Hug 'n Talk Doll and See 'N Say Talking Toy and the "Electronic Speak and Spell that talks." The style section of the *Washington Post* ran a Thanksgiving weekend feature titled "2001 Toys—A Space Odyssey," which highlighted "the phantasmagoric trappings of ultrasonic hyperspace" available, including a sonic ear, ray gun, space fun helmet, lunar patrol playset, and, of course, plenty of *Star Wars* merch. Adults were portrayed as befuddled, while kids were "able to see mechanisms for what they are and not overcomplicate them with abstractions."

Here, Speak & Spell was merely one of a litany of educationally oriented gadgets: Alphie, 2-XL, Mathemagician, Mr. Mus-I-Cal, etc. Its speech feature wasn't even mentioned. The *Los Angeles Times* reprinted a version of the same article a week later titled "Toys That Go Whirrr in the Night." The *Boston Globe* ran its own feature: "Pushbutton Play, the All-Pervasive Computer Invades the World of Kids." TI features prominently, but for its microprocessor (though it's never labeled in that way), not for its speech synthesis. Speak & Spell is described as having a "voice in the computer," but the article quotes a TI spokesperson as saying that Parker Brothers' Merlin "is one of the most technically advanced toys on the market." TI engineers had managed to design Merlin's chips to play back forty-eight musical notes at a time—quite a feat of affordable stored memory in 1978—but there was no difference presented in the article between the different kinds of "silicon chips," nor between Speak & Spell's speech and the bleeping of other games. Many articles did not distinguish between Speak & Spell's speech feature and the sounds coming from other toys. For example, *Newsweek* called Speak & Spell "a challenging voice-command game with a 230-word vocabulary stored inside."[66] There were other talking games on the market, such as Mattel's handheld Talking Monday Night Football, which played sections from broadcast recordings. Neither the average consumer nor the average journalist was aware of, or seemed interested in, the differences in the "black boxes" of these different types of speech technologies.

Electronics continued to "ignite the 1979 American Toy Fair in a blaze of light and sound," but the next year, with consumers still suffering through a significant economic recession, prices had to be dropped to entice parents. During the 1980 Christmas season, Simon, Merlin, and Alphie, all priced in the $20–25 range, were still strong sellers, but TI reduced the Speak & Spell's price from $70 to $50 to maintain sales, even in affluent markets. Some retailers felt the electronics boom had already peaked, seeing that buyers were increasingly picky about electronics, especially handheld games, even if things that were perceived as "educational" were still moving. The toy and game companies invested in convincing parents of the value of educational electronics, separating them from the more dubious value of arcade-type games.[67]

Education is the prototypical activity of information transfer. Although the Speak & Spell seems to have been more of a coincidental than strategic target for voice interface, it was also a logical one. The groundwork for educational technology had been laid by behaviorist ideas of automated and individualized instruction. By the 1970s, education was subject to the same technocratic ideology that other federal services were, and precomputational educational technology was big business, as Audrey Watters detailed in her history of behaviorist teaching machines.[68] Watters showed that those who had influence in guiding national and state-level policies for educational practice, as well as key theorists, were already sold on achieving efficiency in education through mechanization. This had carried over from a Machine Age commitment to scientific management, believing that similar efficiencies could be found for education, where turn-of-the-century immigration required more funding for public schools than many politicians were inclined to provide. Arguments for the classroom use of textbooks, flashcards, and radio were based on industrial logic, as the dean of the School of Education at Stanford University explained in 1916: "Our schools are, in a sense, factories, in which the raw products (children) are to be shaped and fashioned into products to meet the various demands of life. The specifications for manufacturing come from the demands of twentieth-century civilization and it is the business of the school to build its pupils according to the specifications laid down."[69]

Like the Eames/IBM information machine, the right educational equipment would "leave the teacher free for her most important work, for developing in her pupils . . . high ideals," promised the inventor of a 1920s mechanical device for taking, and automatically scoring, multiple-choice tests.[70] In the 1930s, IBM itself worked on electronic equipment for automatic test scoring and explored partnering with the famous behaviorist B. F. Skinner to manufacture a teaching machine of his design in the 1950s.

As Watters stated, "It is impossible to tell the story of teaching machines without telling the story of the business of educational technology."[71] Although John Dewey's ideas about progressive, child-centered education were also popular at midcentury, the Cold War and space race called into question the goals of the US public education system, and industry advocated technological solutions. A 1960 magazine article explained:

> At present, the US is spending an estimated $15 billion annually to prepare 42 million children for the rigors of life in the space age. Many Americans, however, are convinced that this effort is woefully inadequate. While Russian schools . . . are grinding out scientists and technicians like sausages, the vast American educational plant is so overcrowded in many areas that children only attend on a part-time basis. . . . [W]hat amounts to a new industry, devoted to the manufacture and sale of educational devices, is arising. It makes thousands of gadgets for school laboratories, turns out school furniture, puts lessons on film and tape, supplies components for educational television systems and develops such specialized units as the 'electronic learning environment.'[72]

The nexus of power over educational decision making became entrenched with those who advocated technological solutions. As Watters observed, "Major funders of the push for educational technology—the Ford Foundation in particular—largely bypassed teachers," and, consequently, educational expertise shifted from educators to technologists.[73] In the 1970s, the home became a market for educational technologies, with the message that parents didn't need the expertise of a teacher to provide their children a quality experience, but only a technology like Speak & Spell.

Tiger Electronics hired the head of child development in the pediatrics department of the University of Chicago Medical School as its spokesperson to "[inform] consumers of the growing importance of computer toys as valuable educational tools." Appearing on radio and television in several key markets, she highlighted products including Hippo-Lot-O-Fun, "one of the first computers for the crib."[74] In 1979, Harvard education professor Howard Gardner sang the praises of the new crop of computer toys in an article in *Psychology Today*. Gardner, likely eager to distance himself from Skinner, who had recently retired from Harvard, argued that the new "learning toys" were a significant improvement on the "teaching machines" that Skinner advocated in the 1950s. This is ironic, given that we already know that the Speak & Spell engineers cited just those teaching machines in their patents. Gardner spent most of the article describing several toys—Speak & Spell,

2XL, Merlin (his personal favorite), Chess Challenger, and Play 'n' Playback Organ. He argued that what largely differentiated these *learning* toys from the previous era of *teaching* machines was the level of engagement that multiple audiences of children and adults experienced when using them. He said that 2XL had "personality" and you could "communicate *with* him/her/it."[75] It makes much less difference that Gardner chose not to specify a single pronoun for the toy than his easy use of the preposition "with." Since the only input that the machine could respond to was the pushing of one of four buttons, communicating *with* is here defined by a single physical interaction, although presumably Gardner was taking into account the mental process of engaging with the quiz question posed by the toy and calling this communication as well. It's unclear how this algorithm actually differs from that of the behaviorist teaching machines, which also posed questions, waited for physical input, and judged that input as representing a correct or incorrect answer. One key difference was sound—the voices of 2XL and Speak & Spell and the musical tones of Merlin, Simon, and others. Gardner said of Speak & Spell, "The human sounding voice . . . reminds one of a well-meaning and empathetic school teacher," making emphatic the point by elaborating that "one converses with a reliable, if somewhat stilted and stodgy teacher, not a stupid—or omniscient—machine," again choosing the preposition "with," although the tiny number of repetitive responses that Speak & Spell could provide hardly counts as either omniscient or conversant.[76] Other reasons that Gardner gave for the increased engagement with the toys are mostly superficial: their colorfulness and portability, the "space-age sheen" of their robotic voices and command panel keyboards, and the speed with which they responded. By this logic, human teachers are nothing but slow, old-fashioned, even if well-meaning, teaching machines, now outmoded by the new, engaging educational technologies.

Not everyone was enamored of the new electronic toys and games. For example, Paula Smith, a child psychologist employed by Fisher-Price, argued in the *Journal of Children in Contemporary Society* in 1981 that "we must insure that we do not create toys which are brighter than the children whom they are supposed to entertain," an interesting anxiety that took the computer/brain metaphor at face value.[77] Smith wondered, "How do we insure the age appropriateness of a 'mind' toy when we are not even sure how the mind works?"[78] Much as Gardner had, Smith described the various computerized toys that were on the market—toys that made music, video games (still referred to as "TV games" by Smith), arcade games and their handheld equivalents, and games of speed (this is the category that Smith assigned to games like Simon). Smith was most concerned about the physical and mental *passivity* of the child when playing with these toys and games, explaining: "We play on the game's terms. . . . We use eye-thumb or eye-finger coordination. . . . [Speed settings] are present. . . . There is rarely a second chance. The player rarely wins. . . . There is no mastery. . . . Age appropriateness is questionable at best, both due to developmental level and skill emphasis. . . . They are two dimensional illusions."[79]

She further worried about isolationism, a lack of problem-solving and imagination, and a loss of "the real"—real musical instruments, real bats and balls, or real skill in arithmetic. Kids such as those interviewed by *Newsweek* liked the option of being able to play with their electronic games by themselves, giving credence to what some were detecting as "lonely child syndrome" in the 1970s due to low birth rates, more only children, more years between siblings, and two working parents in many families. Consequently, games that could be played alone were of interest to many toy companies. Joseph Weizenbaum, a computer scientist at the Massachusetts Institute of Technology (MIT) who became famously disturbed by people's inclination to believe that his 1960s-era chatbot program Eliza was "smarter" than it was, was opposed to electronic games, which he thought "only embed the child's loneliness. What that child needs is the companionship of other people, not a computer."[80]

But for every critic, there was a booster. In a brief piece in *Video Review* in 1983, Isaac Asimov praised the move toward educational gaming systems. Starting from the assumption that "games in youth are a rehearsal of life in maturity," he first granted that most games might only offer combativeness and eye-hand coordination, but he rebounded by suggesting that more games could be designed like computer chess—educational games that helped apprentice players learn from the computer master. His fictional example was a writing game that he called the "Game of Author": "In fact, you may learn good writing more quickly and more pleasurably than you possibly could in the conventional way of attending a school that offers courses in the subject."[81] Asimov, too, seemed to expect the artificial intelligence (AI) achieved through computation to outpace human creative abilities but also be desirable for social interactions, as the computer game–as-teacher is even more integral to his hypothetical than the computer game–as-author.

MIT anthropologist Sherry Turkle was curious about how young children themselves thought about these new "thinking" toys and games being developed for them, so she asked them. She concluded that children were inventing new psychological categories to describe their interactions with a speaking device like Speak & Spell.[82] Turkle found that children between the ages of four and seven often reasoned that Speak & Spell was alive because it could talk. What was most interesting to Turkle, however, was that when children argued about whether Speak & Spell was alive or not, they used psychological reasoning to support both cases. "It talks, but it's not really thinking of what it's saying," said eight-year-old Adam, to which five-year-old Lucy replied, "You can't talk if you don't think; that's why babies can't talk. They don't know how to think good enough yet."[83] Even older children, who recognized that Speak & Spell wasn't alive in the way that animals and plants are, tended to reserve a category of liminality for computers in general as "sort of alive," especially those that exhibited human characteristics like speech. According to Turkle, these children started to wonder that perhaps reproduction and respiration might not be required characteristics of a living thing after all.

Even though they represent a spectrum of positive and negative responses to these gadgets in children's lives, Gardner, Smith, Asimov, and Turkle all grant the metaphor, if not the reality, of the logic circuit functioning like the human *mind*. Interestingly, they all chose the more metaphysical term "mind" over "brain" when writing about these toys. It was also becoming common in the press to refer to microprocessor chips themselves as "brains," while referring to games (the programmed bit) as having "minds." Human labor was removed from the creation of computer "minds" as surely as it was previously from the design of electronic "brains," leaving the machine framed as an autonomous agent whose role in human childhood was being negotiated by these experts, but also was reflected in a child's use of the toys and games themselves.

While the childhood impulse to animate objects in the world wasn't new to computational objects, Turkle believed that the sophisticated psychologizing was. She explained, "The arguments children use most frequently to discuss whether a computer is alive do not refer to the computer as a physical entity but to . . . ways in which it seems or does not seem to be like a human being in the qualities of 'mind'," including talking and consciousness, intelligence, feelings, and morality. So, for example, Speak & Spell might be alive because "it spells better than me" or "it cheats." Turkle's studies also found that the more contact that children had with computational objects, the more nuanced the child's psychological language became. She observed that computer users weren't just maintaining a willing suspension of disbelief—they actually had to organize new categories for what this thinking/speaking thing was. She claimed this was evidence of the development of a *social relationship* with computational objects.

In the particulars, I don't believe that Turkle was observing a new response to computational technologies or to machinery with lifelike capabilities. From Wolfgang von Kempelen's communicating chess player, to the Voder, to UNIVAC, and the IBM information machines that followed, humans used psychological categories to explain computational phenomena and projected human sociality and personality onto inanimate but communicative machinery. Recall, for example, the Voder described as "talking a blue streak," a description that projects human sociality on the machine.[84] At the beginning of the personal computer era, Turkle was interrogating human relationships with digital machines for what they might tell us about where we stand in the world of artifact, but describing a longer history of this negotiation than perhaps she was aware of: "We search for a link between who we are and what we have made, between who we are and what we might create, between who we are and what, through our intimacy with our own creations, we might become."[85]

Significantly, though, she was among the first to look at how this negotiation played out in the always vexed boundaries of childhood, which is itself historically symbolic of a society's most conservative values. While childhood animism itself is not problematic, animizing the machine struck some as troublesome. Turkle documented parents who worried about their children taking a talking machine like Speak & Spell to bed instead of a book or stuffed animal—parents who were concerned about their children anthropomorphizing machines when they held no such qualms about anthropomorphizing other toys. "Do you suppose she thinks that people are machines?" one mother worried.[86]

As Turkle and others have shown, "mind jargon" goes both ways. As we talk about machines with minds, we also talk about our minds as machines.[87] For children in the 1970s, this tendency might have been naturalized by the growing ubiquity of electronic games, toys, and "learning aids" like Speak & Spell. Rather than experiencing social automation as a conflicted process involving the potential loss of livelihoods, or an existential threat from a "superior" intelligence, or even as just an inconvenient adaptation to daily life, as some adults had previously, children in the 1970s were the first generation that could choose talking machines as playthings and not know anything different. These beeping and talking electronic toys had significant limitations in their actual conversational abilities, but when reinforced by a media environment that consistently included talking computers as friendly and helpful companions, Gen X was the first to fully embrace this possibility.

COMPUTERS ARE PEOPLE, TOO!

In 1982, the Walt Disney Company produced a one-hour television documentary called *Computers Are People, Too!* to promote the release of its new movie *Tron,* one of the first Hollywood films to feature computer graphics.[88] In the documentary, actress Elaine Joyce, playing a skeptic who is afraid of losing her job as an entertainer to media made wholly by computers, has her anxiety soothed by a "computer of the future, Tele-communicative Operative Memory, TOM for short, a computer that talks to you and lets you read his mind" and that functions "as an extension of a person's intelligence"— that teaches when used by teachers and is artistic when used by artists, the embodiment of the Eames/IBM information machine ideal. This was a very different story about computers than that of *Tron,* about a computer programmer who gets sucked into a mainframe computer where he must battle to escape a malevolent operating system. Instead, TOM told a story about people like *Tron's* human creators who were *collaborating* with computers. The message was that people were *interacting* with computers in the same ways, and for the same reasons, that they interact with other people, and giving those computers voices was key to this interaction, at least on television.

TOM was voiced by a human actor, of course, but experts interviewed within the documentary promised both voice and graphics-rich interactions within the decade, and footage showed children using voice synthesis applications on machines at a computer camp. When *Computers Are People, Too!* was shown on television, it was sponsored by Atari and interspersed with commercials for the Atari 400 ("We brought the computer age home!"), in which a young boy learns to speak some conversational French by practicing with a voice-enabled computer program.

When *Computers Are People, Too!* returned, experts explained that the manufacturing economy was gone and that information was now key, and children needed to learn to work with computers to be successful in the future. TOM reassured Elaine, and parents everywhere, that the kids were all right because "video games are their window into our world." The future belonged to computers, so they were being elevated to the category of "people, too." This set up another Atari commercial for its video computer system that featured an expanding group of enthusiastic players crowding onto the family room couch—kids, their parents, grandparents, friends next door, the mail carrier—all partaking in wholesome, training-for-the-future, video game fun. *Tron* was just good guy versus bad guy fiction. There was no reason to be anxious about computers, except to be anxious about *not* having one: computers are people, too!

Other media confirmed it. In May 1982, *TIME* magazine featured the Computer Generation on its cover, and in January 1983, it announced the personal computer as its Machine of the Year, in place of its annual naming of a Man of the Year.[89] In spite of the fact that human body metaphors have been perennially used to describe human-built machinery, and people had been referred to as computers literally and figuratively at previous points in history ("computer" was a job title before it referred to a machine), it still seems a little over the top to claim such enthusiastic parity between electronic computers and people, especially as the title of a family-oriented documentary. But it seemed that the cybernetic metaphor was now a complete analogy, if not equivalency—at least in the hyperbole of the mass media.

The transition from imagining computers as logical brains to imagining them as friendly, social information machines was not abrupt or absolute—a spectrum of cultural ideas about computers existed and continue to exist—but the tendency to align them as friend or foe based on their social performances rather than their logical ones becomes characteristic of family and children's entertainment in the 1970s and beyond. The voices of these computer characters are the interface through which audiences know whose "side" a computer is on. This is seen in animation, given a boost by the development of digital production tools and distribution through cable television, where boy geniuses like Dexter of the animated children's series *Dexter's Laboratory* turned the evil genius trope on its head, making the vaguely European-accented mad scientist with a talking computer someone's annoying little brother, while the multicultural *Codename: Kids Next Door* had a talking computer in a treehouse that helped kids battle adult tyranny all over the world. In the multispecies world of animation, dogs like Ruff Ruffman (*Fetch! With Ruff Ruffman*), with his Fetch 3000, and Courage (*Courage the Cowardly Dog*), with Miriam's abandoned desktop personal computer, also interacted with talking computers. In the undersea world of *SpongeBob SquarePants*, nemesis Plankton married a talking computer. The popular *Knight Rider* series of the 1980s featured the crime-fighting duo of the actor David Hasselhoff and a talking smart Trans Am called KITT (voiced by the distinguished-sounding William Daniels), a computerized, Americanized upgrade of James Bond's Swiss army knife Aston Martin. Superheroes like Tony Stark often had talking AI partners like Stark's JARVIS, while evil computers were parodied as villains, such as D.A.V.E. (for Digitally Advanced Villain Emulator) in the animated *The Batman* series of 2004. Even the Care Bears, in the direct-to-video *Journey to Joke-a-lot* (2004), aimed at a preschool demographic, encountered PAL, a HAL parody complete with red, glowing surveillance camera "eye" and dialogue that repeated, "I'm sorry I can't do that, Dave," when a rat named Sir Jokealot ("I'm not Dave!" he kept screaming) asked it to "open the pod bay doors" to a vault containing a magic scepter.

Even as some adults (most notably US president Ronald Reagan) fretted over the idea of teenaged computer hackers as depicted in television's *Whiz Kids* (1983–1984) and Hollywood's *WarGames* (1983), the idea that the talking computer was the hacker's friend and not just their tool was shown in Richie's relationship with his computer Ralf, which used an actual Votrax speech synthesizer, and David's elaborate computer setup, which also included a speech synthesizer, and which he talked to as he typed commands. Even the war game simulator that David accidentally runs on the NORAD computer has the name of its programmer's deceased son as its password, a way to characterize the computer enemy as an innocent, game-playing child that can be taught human values.[90] In the popular media of Gen X childhood, and that which would follow, kids and talking computers were more often in cahoots than at odds. Through toys, games, television, movies, and computers at home and at school, children of the 1970s were taught that talking machines were people, too. Or close enough, anyway.

5 PERFECT PAUL (1979)

Computers were not, of course, literal people, but people were finding that they were beginning to live in a human-machine hybrid, or *cyborg*, world.[1] Ideas about human-machine bodies on Earth, rather than in space, filled New Wave and, later, cyberpunk science-fiction texts, and even academic discourse, when the biologist and cultural critic Donna Haraway's "Manifesto for Cyborgs" was published in 1985.[2]

Cyborgs weren't just in the movies. The boundaries between human beings and electronic computers were indeed blurring in a very literal sense in the 1980s. Computer enthusiasts reported to the anthropologist Sherry Turkle that they felt they were encountering other minds when they interacted with computers.[3] Multiuser domains and subscription bulletin board services were facilitating real-time, networked communication between people that was mediated by computer—the first blushes of the utopian cyberspace visions that would become ubiquitous with the World Wide Web in the 1990s. Across manufacturing, microprocessors were driving an explosion of industrial robotics, and the first instance of a robot killing a human, at a Ford Motor Company factory in Michigan, made headlines in 1979.[4] On the frontiers of medicine, doctors at the University of Utah School of Medicine conducted the second artificial heart transplant in December 1982, and separate research teams in the US, Australia, and Austria developed electronic cochlear implants in the late 1970s. While *The Six Million Dollar Man* and *The Bionic Woman* were popular on television, numerous researchers pursued real "bionic" solutions to many diseases and disabilities. A bionic future blurred the lines between human and machine even further than the "electronic brain" metaphor had, and more literally. The cultural historian Megan Prelinger has explained that bionics "engages the human body as a place where the two domains of human and [electronic] machine are at work adjusting each other's very definition."[5] Voices, those of people and those developed for machines, were once again sites of these ongoing negotiations.

As smaller and more flexible computer systems came online in the 1970s, especially within educational networks like the Michigan Educational Research Information Triad (MERIT) and the Minnesota Educational Computing Consortium (MECC), computer data transfer by telephone resulted in experimental applications of voice synthesis for educational purposes, as well as for accessibility. A public access news program recorded the "first pizza being ordered by the first computer" at Michigan State on December 4, 1974, when Donald Sherman, who had a condition that caused facial paralysis that made him unable to speak, used a CDC 6500 mainframe computer nicknamed "Alexander," together with a Votrax voice synthesizer over a modem, to call a restaurant called Mr. Mike's and order a large pepperoni and mushroom pizza.[6] It took five tries before Sherman and Alexander could get a pizza place to stay on the line long enough to complete the order. Between the Votrax voice and the long pauses necessary for Sherman to type his side of the conversation, the pizza employees probably thought that they were being pranked and so they hung up. In the fifth call, which was successful, Sherman (via Votrax text-to-speech) began with the statement, "I am using a special device to help me to communicate," and yet the computer was the one described as ordering the pizza in the press coverage, then and now.[7] Like the New Yorker's ambivalence about "mechanized man or humanized machine" in describing IBM's Voice Answer Back (VAB) in 1965, a machine with a Votrax voice synthesizer and a human being without use of his body's vocal system are collapsed together in the act of ordering a pizza over the telephone. Where the human ends and the machine begins is up for debate in the blurred boundaries of communication.

The kind of text-to-speech application that allowed Donald Sherman to order a pizza over the phone had been several decades in the making. Throughout the 1950s, 1960s, and 1970s, federal grants sustained lines of voice synthesis research specifically toward the goal of a text-to-speech reading machine for the blind. This was an important source of funding for voice synthesis in the 1960s, when federal research dollars for electronics and computing were mostly channeled toward aerospace and defense. The communications technology historian and disability scholar Mara Mills has described how disability was used as a pretext for research funding for audial technologies throughout the twentieth century, naming this form of turning disability itself into capital the *assistive pretext*.[8] This economic dynamic was also used during the development of voice synthesis. In the early 1980s, one of the longest running of these reading machine projects came to commercial fruition through technology transfer partnerships between the Massachusetts Institute of Technology (MIT) and two manufacturers, Telesensory Systems, Inc. (TSI) and Digital Equipment Corporation (DEC).

This chapter summarizes the efforts of scientists and engineers over more than thirty years to develop voice synthesis for a text-to-speech reading machine, research that intersected with that described in previous chapters, but that also resulted in applications that were specifically meant to act not as interface but as *prosthesis*. This "prosthetic imagination" gives us additional perspectives on the relationship between voice synthesis technologies and human beings that further considers negotiations between body and machine through experiences of disability.[9] One of the applications of this research, the synthesized voice called "Perfect Paul," created by MIT research scientist Dennis Klatt, ended up being used by the famous astrophysicist Stephen Hawking and became "Hawking's voice," both to himself and the public. Section two of this chapter looks at Hawking's relationship to the sound of the voice synthesizer that he began using in the 1980s.

Another experience of embodiment through which human beings negotiate the world is gender identity. So far in this history, voice synthesizers have been pitched in a range associated with the Western male gender identity, in part due to technical constraints. Some women researchers were determined to change this as computing resources expanded. This prompted questions about how and why gendered synthesized voices were chosen for certain applications. The emergence of numerous voice-enabled electronics throughout the 1980s and 1990s made talking machines increasingly mundane, with benefits, such as safety, in some cases, but also caused concerns about reinforcing the damaging biases that can be rooted in voice stereotypes. This chapter shows both benefits and harms that follow when synthesized voices take the place of human ones.

SYNTHESIS-BY-RULE AT MIT

There is a long tradition of understanding technology as primarily extending the abilities of human beings, a teleological narrative that computers were assimilated into when they were characterized as "brains" extending human cognitive abilities. A closely related idea is that technology can actually improve human beings by making up for the body's deficiencies. In combination, this results in a philosophy of technology that understands all technologies as prostheses and a transhuman ideal future in which human beings leave behind human bodies altogether.[10] On a more realistic and practical level, though, there is a long history of developing technologies specifically to overcome perceived limitations of the human body, or *assistive* technologies, as they've been called since World War II. Alexander Graham Bell's development of the telephone was an outcome of his search for technological "ears" to hear for persons who were deaf, and throughout the twentieth century, sound was explored as a way to allow persons who were blind to "see" printed information.

Extending from the hub of Bell Labs and its speech research teams, academic researchers often took as their goal applications of voice synthesis that were unrelated to efficiencies in telecommunications. Of course, all research needs its benefactors, so speech researchers at the not-for-profit Haskins Laboratories, a team at MIT, and, later, a group at a Stanford University commercial spin-off, TSI, were working with a tradition of federal funding first established by Vannevar Bush during World War II for the purpose of developing a reading machine for people who are blind. Reading-machine research didn't start out with the aim of using voice synthesis to "read" text out loud (text-to-speech), but Homer Dudley suggested that possibility to a meeting of people working on reading machines in the early 1950s, and there was interest in pursuing it. The Veterans Administration (VA)'s Prosthetic and Sensory Aids Service, the primary source of federal funding at this time, earmarked $3 million over thirty years toward the development of reading machines and electronic guidance devices for veterans who were blind.[11] Since both electronic voice synthesis and optical character recognition were still in their infancy at the time and the VA needed to outline a practical program, it agreed to fund three phases of research beginning in 1957: first, the short-range development and testing of an improved optophone, a device being developed at RCA that converted light into acoustic tones;[12] second, two midrange projects to develop machines that could recognize printed characters and generate a spelled (letter-by-letter) acoustic output; and third, a long-range project assigned to Haskins Laboratories to test the usefulness of speech output created by splicing together standardized voice recordings of words to form sentences, as well as continued speech perception research.[13] The speech research community made progress in identifying rules for assigning speech sounds to letter combinations, efforts made easier with the growing availability of digital computing in the late 1960s.[14]

Haskins's funding fluctuated with the vagaries of congressional budgeting, but by 1973 they were planning a Library Service Center with the VA to produce books on demand recorded by machine. Haskins cofounder Franklin S. Cooper described the synthesized speech that could be achieved at this time as "reasonably intelligible and acceptable . . . despite its machine accent."[15] The few user studies they'd been able to perform led them to conclude that the synthesized speech "was good enough for easy comprehension of simple, straightforward materials, but that listening to it put a heavy load on the comprehension of more complex (textbook) materials." In addition, the synthesis rules that had been defined didn't always cover idiosyncrasies in the pronunciation of specialized vocabularies in some advanced materials. Their colleagues at MIT were working on a more complete set of pronunciation rules that they believed would solve this problem.

MIT also had an active reading machine project ongoing since 1957, when the MIT acoustics laboratory disbanded and some of the speech research work was transferred to the Research Laboratory of Electronics (RLE) under the banner of a Speech Communication research group, the goal of which was to understand "the process whereby human listeners decode an acoustic speech signal into a sequence of discreet linguistic symbols . . . and the process whereby human talkers encode a sequence of discrete linguistic symbols into an acoustic signal."[16] Jonathan Allen assumed leadership of the project in 1970. Allen had earned his PhD from MIT with a specialty in speech synthesis and had spent four years at Bell Labs working on the design of semiautomatic information and vocoder systems. He became an assistant professor in electrical engineering in 1968 and inherited the reading machine project soon after.[17]

The team under Allen was focused on the ultimate goal of producing high-quality speech from unrestricted English text.[18] This MITalk project involved creating a comprehensive list of pronunciation rules that then could be modeled mathematically to digitally manipulate audio signals to result in recognizable English words. The MITalk system included rules for syntactical analysis, rules for prosody, and algorithms for producing the output waveform. It involved the work of several researchers over several years, including Dennis Klatt, who created the phonemic synthesis model. The voice created for the MITalk system, "Perfect Paul," was synthesized from Klatt's recordings of himself.

Klatt spent his career at MIT, joining the speech communications group after earning a PhD in communication science from the University of Michigan in 1964. The Bayh-Dole Act of 1980 made it possible for universities, nonprofit research institutions, and individual researchers to own, patent, and commercialize inventions developed with federal funding. MIT encouraged technology transfer and in 1980, Klatt established a corporation to sell a FORTRAN version of the software that he was calling Klatt-Talk Model KT-1. KT-1 incorporated several improvements to the MITalk system and was streamlined to be implementable on a 16-bit microcomputer plus a synthesizer chip.[19] In documenting the product, Klatt explained, "The input format for a sentence to be synthesized consists of a string of phonemes plus phrase and clause boundary symbols. Anyone can learn to use this symbol inventory with less than an hour's training."[20] Although this was true, it was complex for an average new computer user, and Klatt-Talk competed with plug-and-play voice synthesis peripherals like TI's for its own TI-99 series home computers. Those peripherals did not offer text-to-speech, however, and Klatt's company would even prepare phonemic transcriptions for a fee. In outlining the business case for the software, Klatt echoed other boosters of voice interface technologies, but he differentiated his higher-quality system:

> Speech synthesis devices will provide voices for the computer systems of the future. The revolution is upon us as far as vocal response technology is concerned. With the advent of Texas Instrument's Speak and Spell toy, cheap moderately intelligible canned message sets of 200 seconds or more are commercially possible. Five dollar audio response units will soon find their way into automobile warning systems, toys, ovens that talk, and numerous other products and applications where a small set of predictable responses can be preassembled. This waveform coding technology produces natural sounding speech, but the intelligibility is marginal and the flexibility is inadequate for some applications. The alternative proposed here is a more costly system based on high-intelligibility phonemic synthesis by rule. Applications for phonemic synthesizers such as the KT-1 include 1. talking computer terminals, 2. teaching machines, 3. flexible data base inquiry systems, 4. remote access to information over the phone such as banking activity and airline reservations, 5. talking instrument panels for airplanes, 6. hobbyist computers, etc. In conjunction with a text analysis routine, a phonemic synthesizer can be extended to applications such as a reading machine for the blind.[21]

Although a reading machine for the blind had been the original goal of MIT's research, the rapid adoption of personal computers and other consumer electronics in the 1980s had relegated it to almost an afterthought. Indeed, Klatt's list of applications includes all those shown in this book to have been in the market at this time. Klatt Talk was licensed to DEC, a partnership that made sense, given that iterations of the software had been developed on a DEC PDP-9 and written in assembly language. The MITalk version of the software was eventually licensed to TSI.

The MIT/TSI relationship went back to TSI's founding. In 1970, some Stanford professors established TSI to manufacture an optical-to-tactile conversion device for blind persons called the Optacon, which had been developed, in part, through grants from the US Department of Education. James Bliss, an MIT graduate (PhD '61) who had worked with the RLE's Cognitive

Information Processing Group, became TSI's president. Allen and Bliss were in contact, and in 1975, TSI submitted a National Science Foundation (NSF) grant application seeking funding to implement the MITalk speech synthesis software in a TSI text-to-speech product.

TSI's Speech Plus talking calculator had come on the market in 1976 with a twenty-four-word vocabulary using an integrated circuit (IC) speech synthesis chip designed by Forrest S. Mozer and licensed to TSI (Mozer's patent application precedes those of the Speak & Spell group for a similar chip).[22] The company hoped to bring a full reading machine to market but were beaten by Kurzweil Computer Products' Kurzweil Reading Machine (KRM). Given that the market for reading machines was small to begin with, TSI decided not to develop a full reading machine, but it did spin off a subsidiary, Speech Plus, to manufacture text-to-speech conversion products (the conversion of serial ASCII text to speech in real time), including its Prose 2000 board, which was used in the KRM Series 400, released in 1984.[23] The Prose 2000 multibus board, alone, cost $3,500 in 1982, while a speech synthesizer with a stored vocabulary, the Speech 1000, was selling for $2,500 with a power supply. In a letter dated January 22, 1979, Paul Liniak, a blindness products manager for TSI, laid out the company's "Voice Communications Project," calling it "TSI's top engineering priority" to "make available, in a truly cost-effective form, the most comprehensive . . . and natural text-to-speech system in existence—the Modular Speech System developed at the Natural Language Processing Group of MIT, under the direction of Professor Jonathan Allen."[24]

The Speech Plus vision went beyond products for blind persons; it had as its objective "to be the leading independent supplier of speech output products to the information processing industry."[25] Their services included speech encoding using linear predictive coding (LPC) for customized vocabularies that promised flexibility with regard to language, gender, and age as voice characteristics, as well as a TeleWord vocabulary selection service that allowed some customers "convenient and immediate access to thousands of encoded words over the telephone network."[26] Their advertisements explained, "No longer are you limited to display information with meters, lights, buzzers or printouts. Now you can use SPEECH, man's most natural communication medium," echoing their competitors in the market as well as the MITalk developers.[27]

In 1985, Allen, Klatt, and the linguist Sharon Hunnicutt published a book about the MITalk system in which it, like Klatt Talk, had shifted from being primarily about a reading machine for the blind to being primarily about voice interfaces for personal computing. Although a short preface gave the history of the project in terms of its early reading-machine goals, the rest of the book framed the utility of voice synthesis much more broadly. The introduction explained:

> Certainly speech is the fundamental language representation . . . so if there is to be any communication means between the computer and its human users, then speech provides the most broadly useful modality, except for the needs of the deaf, text-based interaction with computers requires typing (and often reading) skills which many potential users do not possess. So if the increasingly ubiquitous computer is to be useful to the largest possible segment of society, interaction with it via natural language, and in particular via speech, is certainly necessary. That is, there is a clear trend over the past 25 years for the computer to bend increasingly to the needs of the user and this accommodation must continue if computers are to serve society at large. . . . It is clear, then, that speech communication with computers is both needed and desirable.[28]

These statements by Allen, Klatt, and Hunnicutt reflect a growing common sense among speech researchers and some computer experts that talking computers were inevitable because speech is the "easiest" and most human-centered interaction that can be had with a machine. This belief is discussed further in chapter 6. In the 1980s, though, the market for talking products, and even voice interface, wasn't as robust as many companies had hoped. For example, TI had originally projected a goal of $34 million in sales of its speech chips for 1981, but by February, it had already dropped that expectation by more than three-quarters, to $8.4 million. Third-quarter projections would end up being half of that.[29] It turned out that the small market for assistive technologies was one where voice synthesis did have an impact. Speech Plus made a significant contribution to the public's ideas about voice synthesis when one of its products was adopted for use by Stephen Hawking.

Dennis Klatt's Perfect Paul became one of the most recognizable synthesized voices ever created when it became Hawking's voice.

SAVING STEPHEN HAWKING'S VOICE

While Hawking's ashes were being interred during a memorial service at Westminster Cathedral in 2018, the European Space Agency (ESA) beamed into space an original piece of music by the Greek composer Vangelis that featured Hawking's voice. The Cebreros station outside of Madrid broadcast the transmission toward the nearest black hole to Earth. ESA's director of science, Gunther Hasinger, said, "It is fascinating and at the same time moving to imagine that Stephen Hawking's voice together with the music by Vangelis will reach the black hole in about 3500 years, where it will be frozen in by the event horizon."[30] The move was meant to honor Hawking's primary contribution to astrophysics, his theory that thermal radiation is spontaneously emitted by black holes due to the steady conversion of quantum vacuum fluctuations into pairs of particles, one of which escapes at infinity while the other is trapped inside the black hole. Hawking radiation, and the related issue of whether information that falls into a black hole is lost or is somehow recoverable from the radiation, was a profound concept that still engenders controversy among theoretical physicists. Interestingly, Hasinger's comments take for granted that the voice made of information, from the voice synthesizer Hawking had used to communicate for more than thirty years, was simply "Stephen Hawking's voice."

Upon his death, many notices in the press made more of Hawking's voice than of Hawking radiation, suggesting he was famous in public for a persona that merged both. Reuters reported that Hawking's synthesized voice "was his tool and his trademark," describing it as a "robotic drawl that somehow enhanced the profound impact of the cosmological secrets he revealed."[31] Even an obituary for Hawking in *Nature* by longtime colleague Martin Rees referred to "the androidal accent that became his trademark" and also speculated that Hawking's field of cosmology was part of what made his life story resonate with a worldwide public—that "the concept of an imprisoned mind roaming the cosmos grabbed people's imagination."[32] Certainly the publishers of Hawking's popular account of the universe, *A Brief History of Time,* must have thought so, as they chose to feature a photograph of a demure-looking Hawking sitting in his wheelchair in front of a night sky full of stars on the front cover of the first US edition of the book.[33] Even if only figuratively, Hawking was one of the world's most famous living cyborgs, moving and speaking by electromechanical means because his flesh-and-blood body could not, traveling the universe in his mind.

Hawking, who had amyotrophic lateral sclerosis (ALS), lost his ability to speak after a trip to Switzerland in 1985 when pneumonia forced doctors to put him on a ventilator and, after being transferred back to the UK, performed a tracheotomy to help him breathe. After recovering from the near-fatal pneumonia, Hawking initially used a spelling card to communicate, indicating letters with a lift of his eyebrows, an interim solution with obvious constraints. Hawking's first voice synthesizer was from Speech Plus, used an iteration of Perfect Paul, and was run on an Apple II computer with modifications that made the system more mobile. It allowed Hawking to communicate at a rate of fifteen spoken words per minute using a hand clicker to select words from a separate software program that displayed them on a screen.[34] ALS is a degenerative nerve disease that affects muscle control, so Hawking's need for an apparatus that allowed him to write was arguably much more critical than the voice synthesis component, but voice synthesis allowed Hawking an additional modality of communication.

In April 1988, Dennis Klatt received a letter from David Maxey, an electrical engineer in IBM's Systems Development Laboratory who was gathering documents to form a speech synthesis history collection for the archives of the Smithsonian National Museum of American History. It said, "Dear Dennis, I heard a National Public Radio broadcast in which the English physicist, Stephen Hawking, gave a speech in the 'Dr. Dennis' voice. I couldn't tell if it was the Prose 2000 or DecTalk. Would you happen to know?" Maxey joked to Klatt, "If Hawking completes the unification theory, we'll give you partial credit!"[35] In a brief handwritten note, Klatt responded, "Thanks for the info on Stephen Hawking. I found out that he has a Prose-2000."[36] Klatt can be forgiven this brevity, as he was suffering from the thyroid cancer that would take his life at the end of that year.

Hawking's public "voice" is certainly created differently than one from a human body's vocal system, but even this voice generated from electricity, signal-processing algorithms, circuitry, software code, and the distant echo of the body of Dennis Klatt became unique in its specific instantiation for Stephen Hawking, even in how it sounded. The synthesized voice became "his" through his body's intimate interactions with and through it. Although the voice itself, as an electronic technology, lacked most of the expressive capabilities of a body's, Hawking was known to use its affordances and constraints as expressions of his personality. He gave his lectures by sending saved text to the speech synthesizer one sentence at a time, but he stated that he did "try out the lecture, and polish it, before I give it," a statement implying that he was practicing for a performance that included revisions to the way that the lecture *sounded* to him as much as to its content.[37] Known for his playfulness, Hawking appeared as himself in or recorded dialogue for numerous television shows, including *The Big Bang Theory* and *Star Trek: The Next Generation*. He recorded the dialogue for his own depiction in animated series *The Simpsons* for three separate episodes and was also made into a *Simpsons* action figure.[38] He told reporters in 2014, "My ideal role would be a baddie in a James Bond film. I think the wheelchair and the computer voice would fit the part."[39]

Hawking's synthesized voice was his "trademark," as obituaries described it, though not literally trademarked by Hawking (it was not his intellectual property). On a web page that Hawking maintained during his lifetime, he credited David Mason of Cambridge Adaptive Communications of putting together the original Speech Plus system for him.[40] "This company manufactures and supplies a variety of products to help people with communication problems express themselves," wrote Hawking. As early as 1988, Speech Plus had offered Hawking a new synthesizer, a CallText 5010, with improved text-to-speech capabilities, but Hawking insisted that the company provide him with one that used the same voice as his previous unit. Of his version of Perfect Paul, Hawking stated, "I use a separate synthesizer, made by Speech Plus. It is the best I have heard, though it gives me an accent that has been described variously as Scandinavian, American, or Scottish."[41] He was known to joke to American audiences that it gave him an American accent.[42]

Hawking famously refused "upgrades" to his communication apparatus, having come to accept the specific configuration of hardware and software, and the sound that it produced, as his own. As the technology that he used to speak became more obsolete, it became easier for Hawking to control other people's use of the synthesized voice that had become so intimately associated with him. Hawking approved of a cut of the biographical film *The Theory of Everything* in 2014 before he let producers use his own synthesizer to rerecord dialogue.[43] The identification of this specific synthesized voice with Hawking went both ways. Not only would many people hear it today and call it "Stephen Hawking's voice" even after his death, but Hawking himself also insisted on keeping the very first synthesized voice he ever used for the remaining thirty-two years of his life, even as voice synthesis technologies improved.

Eventually, the 1980s-era hardware of Hawking's synthesizer was failing, and other factors, from Hawking's continuing nerve degeneration to the incompatibility of his specialized software and input devices with the latest computer systems, made it imperative to upgrade. Several people—graduate students, programmers, and a research team from Intel—were involved in getting a new system working to Hawking's satisfaction. One thing that Hawking insisted on was keeping the version of Perfect Paul that emerged from the specific hardware that he had started using in 1986. The programmer Peter Benie worked on a software emulator of the original microprocessor and digital signal processor of the CallText 5010, without the benefit of the original hardware schematics or software source code. The only thing he had was the machine code.[44] The team struggled to find a copy of the original voice, especially as the company that was once Speech Plus had been sold or acquired four times since. A backup tape from 1986 was finally located by contacting Eric Dorsey, an engineer who had worked on the CallText 5010 and had fielded questions from the press about the synthesizer when Hawking visited California for a three-week lecture tour in 1988. When the team tracked down Dorsey in 2014, he was working for TiVo and was shocked to learn that Hawking was still using the CallText 5010. The new emulator was finally presented for Hawking's approval in January 2018, only two months before he died. He never used the new system in public due to his failing health.

Since first adopting the CallText5010 with its version of Perfect Paul in the mid-1980s, Hawking had many opportunities to upgrade components of his communication system, but he refused because the voice produced by the original setup had become his. Only Klatt's friends and colleagues heard the echo of someone else in Hawking's speech. And they would have heard it all over the place in the late 1980s. Versions of Perfect Paul were a common default for text-to-speech systems, including many "audiotext" phone information systems, for a couple of decades. Even though Klatt's system included other voices like Beautiful Betty and Kit the Kid (based on recorded samples from Klatt's wife and daughter, respectively), Perfect Paul was perceived as the easiest to understand, although it had its limitations. The *New York Times* described Klatt's synthesized voices as "usually understandable but that sound as if they have a foreign accent," higher praise than another system, which the paper described somewhat offensively as sounding "like a scratchy recording of a person with a lisp."[45]

As hardware and voice synthesis software improved over time, the echo of Dennis Klatt grew fainter and the association between Hawking and his specific instantiation of Perfect Paul more solid. "I keep it because I have not heard a voice I like better and because I have identified with it," Hawking explained in 2006.[46] The desire to have a voice that was his own outweighed any benefit from upgrading his systems. In 2013, journalist Lucy Hawking, the daughter of Stephen, interviewed Klatt's daughter, Dr. Laura Fine, on BBC Radio. Lucy was fifteen when her father lost the ability to speak. Laura was eighteen when her father died of cancer. Reflecting twenty-five years later, Lucy said, "[It] means that my father is actually speaking with your father's voice." Laura replied, "[My father] would be so, so thrilled. . . . I had never really thought before how my dad's voice lives on."[47]

Hawking's attachment to Perfect Paul shows how intimate our subjective experience of our own voice is, as well as the relationship that it affords us with the social world, which might reflect our society's biases, as the next section will show. Hawking's use of Perfect Paul conveyed his sense of humor, his intellect, and also his physical disabilities. Even though it was the standard voice on equipment used by many people, it sounded unique as used by Hawking as an individual, to the point that he wouldn't upgrade his equipment because it would have meant changing the sound of *his* voice.

In this example of cyborg collaboration, the machine became an extension of Hawking's body, but one as prone to aging and breakdown as our biological bodies are. Far from transhumanist dreams, Hawking's example shows the experience of being human is a thoroughly embodied one.

BITCHING BETTY

Imagine yourself in the cockpit of a fighter jet, practicing maneuvers over the desert of the American Southwest. Suddenly your altimeter reading is falling and you must act quickly. The complex panel of instruments in front of you should be second nature to use, but in the moment of crisis, the panels blur together, and your body memory must take over. You begin to make adjustments to solve the problem while simultaneously considering the worst-case scenario. A voice interrupts you, firm but calm, in a soothing alto that reminds you of your mother: "Pull up . . . Pull up . . . Pull up," it repeats, and you do what the voice commands, avoiding disaster.

In the 1970s, as McDonnell Douglas was developing the F-15 Eagle fighter jet, testing revealed to engineers that pilots' reactions to warning lights were too slow, especially as the cockpit display increased in complexity. In addition, the development of "heads up" display technology meant that pilots increasingly received information about their aircraft within their field of vision instead of having to look down at a panel of meters and lights. Engineers were concerned that a cacophony of warning bells and buzzers would just add confusion to the mix. One accident report determined that the crew of a high-performance fighter aircraft had "channelized" their attention in an emergency and became unable to prioritize "emergency corrective actions" based on lights and buzzers alone.[48] Testing by the US Air Force had revealed that a verbal warning system would be more effective—that a human voice breaking into the cockpit would convey a sense of urgency, as well as offer clear and unambiguous directions at the point of need. A voice warning system was recommended as "very likely . . . [to] prevent accidents by alerting the crew to take corrective actions instead of remaining fixed on certain aspects of the problem in an emergency."[49] Systems using recorded warnings had already been installed in some aircraft in the 1960s, but voice synthesis promised to make voice warning systems lighter and more reliable. Engineers purportedly chose a female voice for the warnings because they believed it would stand out in an all-male Air Force.[50] A young actress was recruited to record a series of words that were integrated into the warning system of the F-15. That actress, Kim Crow, recalls that after one of the test flights, the pilot was asked how everything worked; he said, "It was wonderful, except for that Bitching Betty."[51] The name stuck.

According to *Green's Dictionary of Slang*, a "Betty," meaning an attractive woman, came into use with reference to long-suffering Stone Age housewife Betty Rubble from the cartoon *The Flintstones* (1960–1966).[52] In the days of recorded warning systems, the B-58 Hustler flight crews referred to that aircraft's warning system as "Sexy Sally."[53] There were also systems that used male voices, the nickname for which was "Barking Bob." Although "Bitching Betty" seems derogatory, some pilots have said that they use it as a term of endearment; the voice warnings can save their lives, after all.[54] "She's insistent, but she's not strident . . . Betty's got a cadence; she's got a snap to her," said the woman who recorded for the F/A-18 voice warning system in the 1990s.[55] It was certainly nothing new in the later part of the twentieth century to refer to a vehicle of transportation in personified and gendered terms, and so hardly surprising that a vehicular voice warning system would be talked about as a person. The new aspect was that the vehicle was equipped with a literal voice, and this voice was a means to provide important information to the operator about the machine's status. As aircraft, automobiles, and other kinds of vehicles incorporated electronics that increased their complexity, operators needed more information about these systems to operate them safely. As the history of Bitching Betty shows, voice became a way to accentuate the most important information, cutting through the cognitive load required to interpret instruments, and providing clear and unambiguous directions. Just as the speed of computation drove the need to get stock traders information at the speed of data processing, the increased complexity of vehicular systems drove the need to categorize and prioritize information in ways that are easy for human operators to respond to. Far from the communication abilities of science-fiction talking computers, these real talking machines spoke in single-word commands and short phrases, but the information that they communicated was often critical. The synthesis of voices that sounded female was new, and as the "Bitching Betty" moniker suggests, voice stereotypes already present in human social relationships were projected onto interactions with talking machines.

Until the 1980s, consumer-grade synthesized voices were in a pitch range that most listeners associated with a male gender. These voices didn't come close to approximating the prosody or timbre of human voices, but they could produce recognizable language and were often identified with the personal pronoun "he." Popular culture often followed technology's lead and had male voice actors perform the voices for talking machines in movies and on television. The higher voices produced by many human bodies assigned female, as well as by children's bodies, required synthesizing other parameters, including additional formants, and consequently required more computing resources. Early attempts at synthesizing female-sounding voices, whether with the Parametric Artificial Talker (PAT) or MITalk, consisted of scaling the formants of the "male" voice, but this did not succeed in "[turning the male voice] into a convincing female speaker," as Dennis Klatt noted.[56] In one paper, Klatt, who studied both speech perception and speech generation, expressed dissatisfaction with the state of research, stating that "it is not inconceivable that the sound spectrograph has had an overall detrimental influence over the last forty years by emphasizing aspects of speech spectra that are probably not direct perceptual cues."[57]

The difficulties were both technical and social. As one technical paper explained, "For high vowels and voiced consonants uttered by women, the first formant is often very close to the fundamental of the voice source spectrum. This makes it harder to measure the first formant accurately." This paper didn't acknowledge limitations with the measurement instrument of spectrographs themselves, but it sited the problem in the female voice, and, by extension, women. It went on to speculate, "The range of possible voices for females is restricted at one extreme by male voices, and at the other extreme by child voices. This implies that listeners will be more critical towards a female synthetic voice than towards a male or even a child synthetic voice."[58] In voice models, female voices were defined largely by pitch, falling between children and men. This was less a technical fact than it was a reflection of structural sexisms that had primarily male researchers considering male bodies as vocal models, and, bizarrely, blaming women's voices for being ill suited to the technology rather than the other way around.[59] In the same way that it was claimed that women's voices were not technically suited to recording and amplification technologies in the early twentieth century, structural sexism limited the scope of some research and technical development that accepted this unsuitability as "natural."

Meanwhile, recordings of female voices providing information and instructions in urban environments—public transportation and security announcements, vending and automatic checkout and teller machines—were increasingly common and chosen to forge what one scholar called a "soft coercion" in the pitch of the neoliberal city.[60] These are voices that tell you where to go, what to do, and how to behave in order to move in an orderly way through the urban environment, and they are meant to maintain calm efficiency, not unlike Bitching Betty.

In November 1983, the *New York Times* published an editorial by sociologist Steven Leveen under the title "Technosexism."[61] Leveen noticed that there were "millions of mechanical objects" now speaking "through the new technology of speech synthesis," including computers, clocks, elevators, automobiles, vending machines, and even bathroom scales. He was concerned that they were perpetuating cultural stereotypes: "These talking machines are associating females with low-level service jobs, while associating males with tasks that are broader in range and higher in status." Leveen identified three "low-status" applications for which female voices were predominately used—supermarket checkout scanners, telephone information machines, and vending machines. That is, female-sounding recorded and synthesized voices had taken over where wage-earning women used to be, as retail clerks, switchboard operators, and cigarette girls.

Leveen had done a little bit of research before writing his editorial. He was aware that synthesizing a higher-pitched voice was actually more "expensive," that it required more data be stored "on a microchip," and he was aware that product developers were willing to absorb that cost because of market research. Most of the "market research" in Leveen's examples amounted to assumptions about gender roles gathered through interviewing mostly professional men. A video game developer: "Have you ever been to a baseball game with a female announcer?" An executive from National Semiconductor: "the [supermarket scanner] systems use exclusively female voices because the male voice . . . sounded 'just a little bit strange.'" Coca-Cola vending distributors (mostly male): "felt the male voice was not as pleasing." And Chrysler, which incorporated a "male" voice into its 1983 cars because testers had stated that when a female voice told them their car's oil pressure was low, it "hit [them] the wrong way." Leveen concluded that "it's not a coincidence that males are usually the ones purchasing the systems, and that they find female voices more desirable," although this preference was domain specific. His concern was that the gendered voice distribution between low-status and higher-status applications would "subtly influence our children's beliefs about which activities and careers are open to them."[62]

Leveen's concerns are often echoed today in criticisms of female-sounding voice assistant applications, although his concerns don't seem to have gained much traction in the 1980s.[63] People were likelier to want to do away with talking machines altogether, as repetitive nagging about needing an oil change or clearing a paper jam is annoying no matter what the voice sounds like. By the time the proliferation of female-sounding voice assistants raised the specter of technosexism widely in the press—Siri, Alexa, Cortana, and Google Now all originally defaulted to female in the US—an evolutionary preference for female voices in informational roles had supposedly been proved by human-computer interaction (HCI) research.

There have now been numerous journalistic articles that have posed some form of the question, "Why are voice assistants always female?" Nearly all of them cite the Stanford communication professor Clifford Nass, whose coauthored book *Wired for Speech: How Voice Activates and Advances the Human-Computer Relationship*, published in 2005, was based on studies that he and his research lab carried out for more than a decade in the 1990s and 2000s. Given his credentials, and with a paucity of other work about human-computer voice interaction available, Nass has been the go-to expert, and remains so, almost a decade after his death. Nass completed a PhD in sociology at Princeton in 1986; there, he worked with James Beniger, whose influential book *The Control Revolution: Technological and Economic Origins of the Information Society*, also published in 1986, argued that the Information Age grew out of a crisis of control in transportation and manufacturing during the latter half of the nineteenth century rather than as a result of the development of electronic communication technologies. Nass's dissertation looked at how information work had changed since World War II, and his collaboration with Beniger explored the application to information work of sociocybernetics, a reversal of earlier "closed-system" cybernetic ideas that instead conceived of individuals as purposive, goal-directed, open systems that continuously generate society through their intercommunication.[64] In the 1990s, Nass's research agenda focused on investigating computers as social actors, which led him to an interest in "interfaces that talked and listened."[65] In addition to directing a lab running experiments on how people reacted to computer applications with voices, Nass was consulting on voice interfaces for Microsoft, IBM, BMW, General Magic, Verizon, and other corporations.[66]

The experiments from Nass's lab were based on a belief that human speech is an evolutionary adaptation and he reasoned that people "behave toward and make attributions about voice systems using the same rules and heuristics they would normally apply to other humans."[67] Although Nass attempted to account for the role of culture in his reasoning about gender and social identity, the experimental designs used in his lab's research could not account for the influence of culture on his test subjects. There is no way to remove a person's lifetime of cultural experience to test the degree to which preferences about gendered voices are hardwired or learned, or reflect some degree of both. Furthermore, in pursuing quantitative measures of people's feelings about voices—how much they like, trust, and find credible a certain synthesized voice—test subjects were typically classified simply as "male" and "female," in a two (gender of voice) by two (gender of participant) experimental design.

In this research design, there is no way to capture the influence of other demographic categories or to ask questions about why a participant might express certain preferences. It is set up to confirm the evolutionary assumption that the human animal's sole motivating purpose is to reproduce, and biological sex is therefore the most important distinguishing category between bodies. Nass demonstrated that bias when he explained, "The gender of a voice—even a clearly machine-like voice—activates the brain's obsessive focus on gender."[68] Nass did acknowledge sexist biases about voices—that people take male voices "more seriously" than female voices, for example—and he ascribed these biases to "learned social behaviors and assumptions," stating as fact that "each culture defines canonical behaviors for females and males."[69] However, his commitment to a biological basis for social behavior forced him into having to make excuses for the sexist stereotypes designed into female-sounding voice interfaces. Nass declared that stereotyping, and dichotomous thinking in general, were the result of people's "limited processing capabilities," and the fewer categories and the greater homogeneity within categories, "the smaller the burden of information gathering and analysis that is required to obtain a given level of certainty and the lower the likelihood of error." If people are just poor information processors (in other words, bad computers), as the computational model of mind insists, then there's no reason to think about human beings developing other capacities, such as greater empathy, since empathy is neither logical nor information based, but, rather, is experiential, relational, emergent, and embodied. Nass concluded that "there is a great deal of evolutionary pressure to rely on gender stereotypes."[70] While finding gender stereotypes "regrettable" and even describing them as "pernicious," Nass nevertheless argued that even if "the human brain fully eliminated gender stereotypes, it would need many more stereotypes, likely just as pernicious, to replace them."[71] His argument relies on a cynical application of unscientific social Darwinism. Meanwhile, *Wired for Speech* recommended the use of these stereotypes in voice interaction systems (male for cars, female for service applications) because the best business case is one based on existing biases.

There were women scientists who believed that solving the problem of technical sexism lay in creating better synthesized female voices and providing plenty of them to choose among. A *New York Times* article in October 2000 put Nass in conversation with the phonetician and speech technology entrepreneur Dr. Caroline Henton, who argued that listeners' prejudices should actually be contested through voice synthesis. The article also quoted several men affiliated with companies in the growing e-commerce market who just wanted to use a voice that had the greatest potential to appeal to customers. "We want to match our experts on the fly to shoppers," explained one. "There may be a dispassionate voice, like Mr. Spock's, for those who sound as though they want to get down to business, for example, and a more nurturing voice for those who seem to want to take their time."[72] In a 1999 article in the *Journal of the International Phonetic Association*, Henton echoed Dennis Klatt in arguing that the lack of quality female-sounding synthesized voices was due to "insufficient data on female speech production" and "inadequacies in analytic hardware."[73] One motivation for better female-sounding voice synthesis was to provide a wider range of options for persons with oral speech disabilities. However, Henton's central concern, like Nass's, was commercial. Since "the lack of appropriate (female or age-related) voices are the most commonly cited objections to not using [text-to-speech]," she reasoned that the future of speech technology was female. Henton would go on to be a phonetician at Apple, working on Siri.

At Bell Labs, Dr. Ann K. Syrdal was also working to create more natural-sounding female voice synthesis. Syrdal's experience reflected some of the most active voice synthesis research in the US, completing her graduate work in psychology at the Center for Cognitive Sciences at the University of Minnesota, and then becoming an affiliate at Haskins Laboratories and spending a five-year NIH grant period at the RLE at MIT before joining the speech technology department at Bell Labs. Her Bell Labs team created what some specialists consider the first high-quality female-sounding synthesized voice, called Julie, which won an international competition in 1998. They accomplished this by focusing on concatenative techniques and the development of algorithms for combining snippets of recorded speech together rather than on programming thousands of syntactic rules and trying to capture all the exceptions to those rules (as Klatt's work had done). Some of this concatenation had been done by a Japanese company from whom AT&T licensed a large database of sound variations. The Bell Labs team sliced those sounds further, making segments shorter than even a phoneme. They found that they could create any number of voices with enough recordings.

The *New York Times* reported on the project, saying that it was the first step toward one day being able to apply a "voice font" to any digital text as easily as one could change its typography.[74] There was no mention of "deep fakes" or other concerns about deception, just the hope that any voice could be synthesized in the future for any reason. The wide deployment of a limited number of voices for a handful of applications was not an issue that researchers considered, much less the perpetuation of damaging stereotypes that this might cause. Henton, Syrdal, and others viewed the lack of knowledge about female-sounding voices and the lack of natural-sounding synthesized female voices as itself a form of technosexism. Scholars including Leveen, and more recently Yolande Strengers and Jenny Kennedy, have argued that female-sounding voice assistants are sexist, perpetuating white, middle-class, heteronormative fantasies about women's compliance with men's needs, as well as gendered labor hierarchies.[75]

In broad strokes, these criticisms are warranted. The corporations that use voice synthesis for commercial purposes are invested in using voices that seem to correlate with the purchasing and other behaviors that they want to encourage. They have often followed Nass's advice in choosing differently gendered voices to match sexist expectations. Female-sounding voices are supposed to be calming and male-sounding voices more authoritative. But over the last several years, there has been a shift toward often younger, male-sounding synthesized voices for many applications, including domestic and customer service assistants: in 2015, the UK grocery chain Tesco changed the voice of all its self-checkout machines from female to male; IBM's Watson modeled the vocal quality of the typical *Jeopardy!* winner—an educated white man in his mid-twenties to forties, and then became a *Jeopardy!* champion itself; Jibo, a social robot for the home, was supposed to be another member of the family, and developers chose a friendly and enthusiastic young adult male voice for it modeled on Michael J. Fox's performance of Marty McFly in the *Back to the Future* films; and Apple offers several voices for Siri, including male- and female-sounding voices with subtle characteristics of African American Vernacular English, and no longer defaults to the original female unless the user chooses it. According to the *Guinness Book of World Records*, the most downloaded sat nav voice before Google Maps became widely used for personal navigation was the animated oaf Homer Simpson, as voiced by Dan Castellaneta.[76]

It is likely that the pervasiveness of voice synthesis now promotes its own stereotypes. For example, the British butler voice assistant in popular culture, including the KITT smart car in *Knight Rider* and Tony Stark's J.A.R.V.I.S. in Marvel's *Iron Man* movies, has class implications. Meta chief executive officer Mark Zuckerberg created his own Jarvis smart home system (extremely rudimentary compared to Stark's fictional one, of course) for controlling lighting, temperature, entertainment systems, and other functions. When he asked the public who should be the voice of his Jarvis, more than 50,000 people suggested the deep-voiced actor Morgan Freeman, whose voice the Netflix Film Club has called "the most calming of all time."[77]

The mandate of informational capitalism includes colonizing the home not only to mark it as commercial space, but also to extract as much informational value from it and its inhabitants as possible, as chapter 6 shows. Networked information and communication technologies are increasingly surveillance technologies. Some of this surveillance is silent, but some of it speaks in the voice of a smart home networked information system and other talking Internet of Things gadgetry. Whether that system speaks in a "smart wife" vocal stereotype or the dulcet tones of Morgan Freeman, giving the system a voice sustains the illusion that corporate informational interactions are interpersonal ones, and interpersonal interactions are primarily informational. Put another way, changing the sound of Siri's voice (something that is easy to do) doesn't change the fact that Siri is the "voice" of a US-based technology corporation that manifests a great deal of power by controlling the information collected and provided through Siri. Using our biases for this purpose is what is important to information corporations, while perpetuating stereotypes is left out of their priorities because they define that as a cultural problem rather than a technological one.

The cultural problem can *be* a technological problem. We learn to value the humanity of people that we perceive as different from ourselves through experience. As synthesized voices become common, replacing with networked technologies what might have previously been interactions with other people, we lose exposure to the vocal diversity and expressiveness of other human beings and risk losing some of our capacity for understanding others accordingly. However, synthesized voices that can more fully simulate human expressiveness are not a solution to this problem. In fact, they would significantly exacerbate it by opening up unlimited possibilities for deceit and manipulation.

VOICEBANKS

One argument for improving the naturalness of synthesized voices has been to provide voice surrogates for people who cannot speak themselves. The voice artificial intelligence (AI) company VocaliD promotes a rhetorically powerful origin story about its founder, Dr. Rupal Patel, being inspired to start the company when she walked into the exhibit hall at an assistive technology conference and heard a young girl and a grown man having a conversation through their respective assistive devices, both in the same Perfect Paul voice. Dr. Patel was "astonished" to recognize that "literally hundreds of individuals [were] using just a handful of the same generic voices. Voices that didn't fit their bodies or personalities."[78] Invoking a false equivalence, Dr. Patel argued that "we wouldn't dream of fitting a little girl with the prosthetic limb of a grown man, why then was it okay to give her the same prosthetic voice as a grown man?"[79] In an inspirational TED Talk, Patel exclaimed that "science is an incredible superpower when applied to improve the human condition."

But, of course, VocaliD's products are not limited to a few hundred custom voices to be individually used and controlled by the persons for whom they are designed. The company's intellectual property for crowdsourcing voices could potentially be used to create voices that could mimic others, although recent tests show that forensic techniques can still differentiate between synthesized and natural recordings of a single voice.[80] Even with as much data as VocaliD collects, synthesizing a voice that can mimic the full spectrum of expressiveness that a single human body can, and use it dynamically in social contexts, has so far proved difficult, if not impossible, but this might not always be the case.

What the history of gendered voice synthesis, text-to-speech technologies for blind persons, and Stephen Hawking's use of Perfect Paul shows is the ongoing negotiation of what it means to be human in a cyborg world.

Voice is symbolic of full participation in the US political economy on one hand, with voice synthesis held up as an ironic dream in which differences in human embodiment are overcome through technological means. On the other hand, an individual's identity with their own means of expression, their own "voice," no matter what it consists of, remains a potent experience of subjectivity and the basis on which we relate to others. The synthesized voices of talking machines, then, are not only an interface through which human bodies can interact with those machines, but also the field of negotiation where human beings come to figure out what the boundaries between human and machine *should* be and how much those boundaries matter to us. User-centered and human-centered are not the same.

6 S.A.M., THE SOFTWARE AUTOMATIC MOUTH (1982)

On January 24, 1984, after considerable media hype, Steve Jobs stood on stage in a packed 2,400-seat auditorium to introduce Apple's new Macintosh personal computer. His sidekick for the presentation was, in fact, the Macintosh itself.

"Hello, I'm Macintosh. It sure is great to get out of that bag," it joked. The crowd erupted in applause and shrieks of surprise and delight.

Preparation for the demo is imagined at the beginning of the 2015 movie *Steve Jobs*. In the film, Jobs (played by Michael Fassbender) argues with his marketing exec, Joanna Hoffman (Kate Winslet), over cutting the voice demo, which he has been told will certainly crash.[1] "You have to tell me why it's so important for it to say 'hello,'" Hoffman demands.

"Hollywood. They made computers scary things," explains Jobs. "See how this reminds you of a friendly face?" he continues, pointing to the Mac, with its wide-screen "eye" and floppy disk slot "mouth."

"It's warm, and it's playful, and it *needs* to say 'hello'!" he screams.

In putting these words in Jobs's mouth, the movie's screenwriter, Aaron Sorkin, certainly had in mind the crew-murdering HAL 9000, perhaps a nod to Jobs's roots in the late 1960s counterculture. Since the debut of *2001* in 1968, HAL's menacingly matter-of-fact statement, "I'm sorry Dave, I'm afraid I can't do that," has transitioned from science-fiction nightmare to iconic pop culture sound bite. The first version of Apple's Siri voice assistant on the iPhone 4S (2011) included this Easter egg: if you said "Siri, open the pod bay doors," Siri's response was, "We intelligent agents will never live that down, apparently."[2]

Apple did not develop voice synthesis technologies on its own, but the visions that the company promoted for easy-to-use, everyman computational devices serves as the backdrop for developments that led voice synthesis, together with speech recognition, to be promoted as the holy grail of human-computer interaction (HCI) in the 1980s and 1990s. From the talking Mac in 1984, to the John Sculley–era vision of the "Knowledge Navigator," a talking information concierge in 1987, to Apple's acquisition of the Siri personal assistant in 2010, Apple was the most high-profile US company to consistently pursue voice interface for its products over these decades, even as a large chunk of the development of the technology itself was funded by the US Defense Advanced Research Projects Agency (DARPA) and parceled out to a large number of university and institutional research teams. While the statistical natural-language processing (NLP) that would make two-way voice interaction with Siri possible was being developed, applications of voice synthesis that required no or only limited speech recognition, or that worked with text-to-speech software, continued to proliferate. In the administration of daily life, many people encountered applications of voice synthesis over the telephone, in their transportation systems, and at the bank and the grocery store. Daily life at home was the next target for automation through synthesized voice-enabled technologies.

This chapter looks at voice synthesis in the 1980s and 1990s as conceptualized by the newly institutionalized discipline of HCI, especially as employed at Apple, and follows the path of voice-enabled devices into middle-class homes as home computers became not just desktop devices in home offices, but part of an information network that helped manage the household, including the people in it. In previous chapters, we've seen how voice, the means of communicating between most human beings, became the way that information machines communicated with people in the popular audiovisual media of television and movies, and by some actual machines themselves. In this way, informational exchanges between people and talking machines took on some of the characteristics of interpersonal exchanges, and primed people to accept voice interface as the next improvement for personal computing technologies. After all, these technologies were coming from an industry that was enjoying the increasingly significant role that it was playing in shaping cultural trends, not to mention financial success, both of which provided companies like Apple with a great deal of political and economic power by the end of the twentieth century.

Jobs, in particular, seemed to encourage his own mythic status, portraying himself and the company he founded as David slaying the giant IBM by bringing user-friendly computers to the masses. Ironically, Jobs brought Charles Eames's *Information Machine* vision for IBM to its ultimate realization in Apple's "Think different" marketing campaign (1997), a direct provocation in response to IBM's longtime slogan, "Think." The campaign's television commercial featured "the crazy ones, the misfits, the rebels, the troublemakers" as it showed pictures of cultural luminaries including Albert Einstein, Bob Dylan, Martin Luther King, Jr., Mahatma Gandhi, John Lennon, Amelia Earhart, Pablo Picasso, and many more people who "push[ed] the human race forward."[3]

Something that wasn't pictured or spoken of was a computer. When Jobs returned to Apple in 1997, he positioned the company as reaching far beyond computers, appending the lowercase "i" to the plethora of digital devices and services it would release: iMac, iPod, iTunes, iPad, and, of course, the iPhone. A computer is now something that a majority of us, of all ages and in most nations, carry with us everywhere we go, even from room to room inside our own living spaces. This chapter shows how domestic spaces have become colonized by digital information devices and the corporations that provide them. As we continue to navigate and debate the use of so-called smart-home technologies, their synthesized voices both lull us into the fantasy of *Jetsons*-like companionable services from autonomous talking machines, but are also flawed enough to remind us that relationships with people, unlike social interactions with machines, are about more than information exchange.

FROM S.A.M. TO SIRI

The public debut of Siri as a native feature of Apple's iOS operating system in 2011 marked a significant development in voice synthesis technologies, but voice interface had been on the drawing board at Apple for thirty years before Siri was released. The Macintosh computer's voice synthesis software was developed by Mark Barton, who had written the first commercially available software program for speech synthesis, S.A.M. (whose initials stand for "the Software Automatic Mouth") as an independent program for the Apple II. S.A.M. appealed to personal computing's early adopters, and, in 1982, it came to the attention of Steve Jobs, who wanted a text-to-speech system for Apple's secret new project, the Macintosh.[4] The Macintosh had originally been conceived of by Apple's manager of publications, Jef Raskin, who had done graduate work in computer science and had a vision for the future of personal computing that included a networked "information appliance" in every home.[5] The journalist Steven Levy has said that it was Raskin who provided the vision for the Macintosh's "groundbreaking friendliness."[6] Even though the machine that Jobs shepherded into production was very different from the one that Raskin had described, the fingerprints of early HCI ideals about "user-friendliness" were one of the Macintosh selling points, and the idea that voice was the most user-friendly interface of all had taken root.

Barton was introduced to voice synthesis as a junior high student in the mid-1960s through a science kit developed and distributed by Bell Telephone Systems. One of several hands-on kits for science classrooms, Bell's Speech Synthesis kit allowed students to create a simple synthesizer using three transistors that could make all five of the long vowel sounds.[7] In high school, Barton and a few friends taught themselves to program in FORTRAN on an IBM 1130 computer, the only computer in the entire Los Angeles city school system for student use, which just happened to be located in his high school. Later, at the University of California at Los Angeles, Barton put together an analog formant synthesizer that was controlled by a BASIC program that he'd written for an Apple II computer that he had bought for that purpose. He wondered whether it would be possible to just have the computer act as the synthesizer, so he started working on writing a software program that could reproduce his analog synthesizer's output. A bit to his surprise, it worked. Barton had written a voice synthesis program in assembly language for the Apple II that was "very, very buzzy, but readily understandable."[8] He ported the program to the Commodore 64 and the Atari 400/800 as well and sold more than 50,000 copies. A significant amount of research went into his program, with Barton studying, in particular, the phonemic rules published by the group at the Massachusetts Institute of Technology (MIT), although S.A.M. was much simpler and only distinguished between "content" words and "function" words as parts of speech, plus punctuation for the purposes of a minimal prosody. S.A.M.'s speech quality was much poorer, but its software was significantly cheaper than the DECtalk hardware.

When Jobs came knocking, Barton recruited a friend from his IBM 1130 days, Joseph Katz, to do all the system integration, and the two completely rewrote S.A.M. to create the software known as Macintalk. Unlike the linear predictive coding (LPC) synthesis of Speak & Spell, the Macintalk voice was generated from scratch through signal processing alone, like the parametric systems that came before, and it could be difficult to understand if one wasn't used to it. Barton programmed the script for the Macintosh demo himself, fine-tuning each sound to be as understandable as possible, and had the text displayed on a large overhead screen as the Macintosh spoke, just to ensure that the audience understood what was said. Barton and Katz formed a company called SoftVoice and reworked the software again for the faster next generation of personal computers, adding digital filtering to improve the speech quality.

Jobs had not required Barton and Katz to keep an exclusive license with Apple, so they also licensed the software to IBM, Hewlett-Packard, and Microsoft, as well as some smaller companies. Texas Instruments (TI) and Atari had speech peripherals for their own home computers. At Apple, Macintalk floundered for a bit after the initial release of the Macintosh before the company invested in some in-house speech engineering that resulted in Macintalk version 2 for System 6 in 1988. By the mid-1980s, there were numerous products on the market for making personal computers talk. Given their fairly limited vocabularies and awkward electronic voices, as well as limited speech recognition capabilities, none of these products rose to the success benchmark of "killer app." Nevertheless, the powerful vision for natural-language communication between computers and people often drove computer speech research, and it certainly drove a number of commercial dreams as well.

Even with Jobs gone from the company, famously forced out by its board in 1985, Apple pursued speech technologies during the tenure of his successor, Sculley. When Siri was released in 2011, several journalists made the connection between Siri and Sculley's vision, as articulated in the epilogue to his 1987 memoir *Odyssey*, for a "twenty-first century . . . wonderful fantasy machine called the Knowledge Navigator, a discoverer of worlds, a tool as galvanizing as the printing press."[9] Sculley described the Knowledge Navigator as a networked multimedia device that used a talking software agent to find and manage customized information for the user (not far from Raskin's ideas for the Macintosh).[10] Sculley was not a technologist, though—he was a marketer, and the features of the Knowledge Navigator represented the ideas of many people in the expanding interdisciplinary area generally called HCI, an iteration of an earlier paradigm that emerged from the language of cybernetics as man-machine communication.

HCI was institutionalized in 1982 with the first ACM Conference on Human Factors in Computing Systems and the formation of the ACM Special Interest Group on Computer-Human Interaction. In a 1989 review paper, the HCI guru Jonathan Grudin provided one history of the evolution of user interface development:

> Initially, the user interface was located at the hardware itself—most users were engineers working directly with the hardware. The focus then moved to the programming task—higher level programming languages and environments progressively freed users from the need to be familiar with the hardware. Next, with the widespread appearance of interactive systems and non-programming "end users," the user interface shifted to the display and keyboard, with early attention to perceptual and motor issues. Recent years have seen increasing research focus on the users' "conversational" dialogues with systems and applications, involving deeper cognitive issues underlying the learning and use of systems: the user interface is extending past the eyes and fingers, into the mind.[11]

"Into the mind" was operationalized through psychological perspectives focused on modeling users' goals and developing interfaces that adapted to and anticipated an individual's information needs.[12] In the name of user experience, voice interface and speech recognition were being developed as the most direct route of interaction between human and computer "minds," although we've seen that this was neither a new idea nor a new avenue of technology development in the late 1980s. In Grudin's history of the discipline, HCI was located at the hardware interface in the 1950s, at the software interface in the 1960s, and at the terminal interface in the 1970s with "end users" (people who were neither engineers nor programmers) only coming to the fore in the 1970s; however, there had been end users of computed information since the 1950s, and computer manufacturers, telecommunications researchers, and the public were invested in what access end users had to this information and how they accessed it. As we've seen, the public was increasingly aware of the power of computational information systems in society and debated the effects of this power through tropes of talking machines in popular media. Interface as "dialogue" was an established conceptual paradigm and had been under development as voice synthesis since the earliest days of commercial computing. Even so, NLP for speech recognition was still limited, and experts disagreed on the timeline for realizing Knowledge Navigator–like functionality, or even if truly conversational interaction was possible.

Sculley's future-oriented epilogue was based on extensive conversations with Alan Kay, who was then an Apple fellow. Kay might not be a household name outside computer science circles, but he is famous within them. He earned a PhD from the University of Utah in 1969, from the pioneering program in computer graphics run by David Evans and Ivan Sutherland, which also graduated the future heads of Adobe and Pixar.[13] As a graduate student, he was involved in research funded by ARPA's Man-Machine Communication initiatives, priorities of the psychologist J. C. R. Licklider, who was head of ARPA's Information Processing Techniques Office (IPTO) from 1962 to 1964. Licklider was dedicated to a cybernetic vision of "man-computer symbiosis," as he outlined in a 1960 paper, and "the computer as a communication device," as he outlined in a 1968 paper; and he was instrumental in funding, through IPTO, the basic research that led to the ARPANET.[14] Often, his association with networking overshadows the fact that Licklider began his career in psychoacoustics studying perceptions of pitch, research that contributed to developments in voice synthesis as well.

Kay joined the Xerox Corporation's new Palo Alto Research Center (PARC) as its Learning Research Group director in 1970, where he was instrumental in designing a graphical user interface (GUI) with overlapping windows famously "poached" by a young Steve Jobs for his new personal computer manufacturing company, and then a young Bill Gates for *his* new software company. Many of the Knowledge Navigator's features are anticipated in Kay and his colleague Adele Goldberg's mid-1970s vision for a Dynabook, "a personal dynamic medium the size of a notebook which could be owned by everyone and could have the power to handle virtually all of its owner's information-related needs."[15]

Kay's description of the Dynabook doesn't include voice interface, as he was developing prototype technologies and the existing prototypes of speech recognition were too limited in the mid-1970s. Nevertheless, Kay and Goldberg's descriptions of the Dynabook echo the conversational analogy of Licklider's work and also are informed by the influence of the mathematician Seymour Papert and his constructivist learning theories about educational technologies.[16] Beyond just "materializing thought" (i.e., recording ideas in static media, like words on a page), Kay and Goldberg's vision was of personal computing as an active "metamedium" that would "respond to queries and experiments," thereby involving the learner "in two-way conversation," interaction that previously had been possible only "through the medium of an individual teacher."[17] This was not the "teacher" that Speak & Spell was. Constructivism is idealistically learner-centric; as Kay and Goldberg explained, the Dynabook kept children in control through the interactive nature of dialogue—they would "have the feeling they are doing *real* things rather than playing with toys or working out 'assigned' problems."[18]

In spite of Kay's role at Apple, and Sculley's appropriation of these ideas for his book's epilogue, products in development at Apple in 1987 were not perceived as reflecting a compelling vision of the future. Sculley needed a new vision to market, especially as he was still competing with Jobs, who had formed a new company and was developing the high-end, higher ed–targeted NeXT computer system. Sculley was invited to give the keynote at the 1987 EduCom, a highly influential educational technology conference and trade show. He needed to take full advantage of the opportunity to bolster Apple's reputation with this group. Sculley planned to highlight some research projects that Apple was funding, but he wanted to follow that up with a vision of the future of computing.

Key members of Apple's marketing and creative services teams met with Sculley about his speech and proposed a "science-fiction video."[19] The marketing team worked with Apple's Advanced Technologies Group to compile a list of emerging technologies to illustrate in the video. These included networked collaboration, shared simulations, intelligent agents, rich multimedia, and hypertext—all ways of dynamically interacting with information, with voice as the interface. Hugh Dubberly, who was part of the team, listed some of the additional sources they consulted while creating the video, a mix that included chats with the noted user-interface design consultant Aaron Marcus and books like Stewart Brand's *The Media Lab: Inventing the Future at MIT* (1987) and William Gibson's cyberpunk debut, *Neuromancer* (1984). Dubberly explained the imperative of depicting voice interface in terms of media production: "The video form suggested the talking agent as a way to advance the 'story' and explain what the professor [in the video] was doing. Without the talking agent, the professor would be silent and pointing mysteriously at a screen. We thought people would immediately understand that the piece was science fiction because the computer agent converses with the professor—something that only happened in *Star Trek* or *Star Wars*."

As with depictions of computers in previous audiovisual media, highlighting a machine in an interesting way almost required that it talk. However, with more than a decade of experience of actual talking machines in the consumer domain, even very limited ones, and coming from a company noted for branding itself as cutting edge and user focused, the Knowledge Navigator video is less "science fiction" than it is a compelling public argument about the near future's promise of personal computing, just as Sculley had intended.[20]

As the video opens, a middle-aged white male professor enters his office and opens a thin, portfoliolike object that beeps and begins talking in an even-tempered and even-tempoed male voice. On the screen, a live-action male avatar, reflecting the ethos of the professorial human user by wearing a bow tie, proceeds to have a conversation with the professor about available scholarly literature in which it becomes clear that the avatar knows who the professor's friends and colleagues are and can apparently read and summarize any article instantaneously, tailored to the professor's specific, unarticulated information desires. The device also produces charts and graphs from data pulled out of articles, and manipulates them according to new parameters as the professor asks questions about them simply by talking to the machine. The device includes no keyboard, stylus, or other input mechanism, although the screen is a touchscreen on which the professor can swipe, drag, and expand displayed items. Everything else is accomplished by voice. The video ends with the professor leaving for lunch and the avatar answering a call from the professor's mother, telling her, "Hello. Professor Bradford is away at the moment. Would you like to leave a message?" An annoyed female voice is then heard saying, "Michael! This is your mother; I know that you're there. I'm just calling to remind you to call your sister . . . ," at which point the video fades out, with the nagging, ignored mother on the line talking to an empty room, a bit of "comical" sexism suggesting that the future of technology will eradicate any need for getting information from nagging women like secretaries, housewives, and even mothers.

Claims that voice interface was the most user-friendly and "natural" way for humans to interact with information machines could disguise subtle sexist, ableist, and racist normative assumptions, as well as the unarticulated, ongoing problem of whose priorities would become embedded in the information controlled at the machine. There's no concern about copyright, citation, verification, and other potential issues with the information that the device finds and manipulates. The professor's ease of use masks the source and selection of the information, which could be controlled by the corporation providing the device and its software. The bow-tied avatar with his friendly voice makes preparing the professor's lecture so pleasant and fast that these other aspects of information use are easily ignored.

It would be more than two decades before so-called expert models of computer-supported cooperative work and statistical language-processing techniques resulted in viable information assistant software, though still far from the Knowledge Navigator scenario. Products that weren't quite smart enough failed, like Microsoft's infamous Clippit, or "Clippy," an anthropomorphized paper-clip avatar for its Office Assistant intelligent user interface for Microsoft Office for Windows 97. When Steve Jobs was restored to Apple after it bought NeXT in 1997, he was determined to succeed in the personal digital device market where Sculley's Newton and Apple spin-off company General Magic had made technical progress but failed to produce a popular product. In 2007, Apple introduced the iPhone, with all the fanfare that the public had come to expect from Jobs's marketing events. Four years later, the iPhone would be made to talk with the addition of the Siri digital voice assistant. Jobs purportedly had paid the makers of S.A.M. somewhere in the six-figure range to make the Macintosh speak.[21] Twenty-seven years later, Apple acquired Siri for more than $150 million.[22]

It was Jobs, again, who had contacted a small group of developers who had something he wanted for Apple's products. Siri's creators conceived of their application as a "do engine" rather than a search engine—a software agent that could connect with forty-two separate web services to return one answer, rather than a list of sources, to a multifaceted question like "Is there a casual Greek restaurant near my afternoon appointment?" The software achieved this by "interpreting" information from one's calendar, restaurant review services, maps, and so forth. It allowed you to "have conversations with the internet," if, by "conversation," you meant to engage in only certain kinds of information queries. Echoing every booster of computer automation from the twentieth century, a journalist in 2013 enthused of Siri that it could eventually be "a human-enhancing and potentially indispensable assistant that could supplement the limitations of our minds and free us from mundane and tedious tasks."[23]

This was certainly the goal of the DARPA-funded artificial intelligence (AI) project, of which Siri was one commercial result. Called the Personalized Assistant that Learns (PAL) project, DARPA described its goal as "developing machine learning technologies to make information understanding and decision-making more effective and efficient for military users."[24] DARPA explained that these software assistants would "reduce the need for large command staffs, thereby enabling smaller, more mobile, less vulnerable command centers." A four-minute promotional video announced the mission to "radically improve computer support to commanders and staff," and told the story of a new analyst, who, on his first day at a new posting, has to deal with a terrorist threat, but he succeeds because he is already trained to use the PAL system. At this command center, PAL "already knows the battle rhythm" and will get its human user "up to date" (rather than the other way around). PAL greets him in a robotic female voice: "Good morning, Major Reeves. I have available current . . . schedules, personnel lists." Major Reeves asks questions of the system and is able to control what information the system focuses on through his natural language voice commands and by selecting parameters by touch on his computer screen. Another human user of the system is shown interacting at his own computer station with a robotic male voice. PAL is personalized not only about the goals of this command center, but for each of the human analysts at work as well, although the system is always referred to as "PAL" and "it" by its human users, reflecting its serious military applications. At the end of the video, a female analyst asks Major Reeves, "Remember what life was like before PAL?" to which he responds, "I try not to." The short, definitive text titles reveal that the exigence for AI is an Information Age version of that for industrial scientific management: "Better, faster decisions. Fewer people." After years of failing to live up to the hype, AI projects were viewed by some military stakeholders as a waste of resources, so PAL project leaders were likely eager to tout its potential benefits in a compelling way, again relying on the audiovisual medium and the illusion of frictionless voice communication to argue for the system's benefits.

DARPA chose the nonprofit scientific research institute SRI International to coordinate the five-year, $150 million PAL project, which included research at more than twenty separate academic and commercial institutions. SRI had been the research home of Douglas Engelbart in the 1960s, when it was still the Stanford Research Institute and where ARPA had funded Engelbart's Augmentation Research Center (ARC) to develop an HCI system that would realize his belief that computers could help solve the planet's most important problems if they could be made to support the organization of a collective human intellect.[25] Progress on this system was demonstrated by Engelbart in 1968 as part of the ACM/IEEE Computer Society's Fall Joint Computer Conference in San Francisco, which included hardware and software innovations like hypertext, word processing, videoconferencing,

GUI, and the computer mouse, among others—a presentation that lives in the annals of computing history as "the mother of all demos." The ARC collapsed in the early 1970s due to a number of factors related to funding, as well as differing opinions about the future of computing and HCI.[26] Some ARC researchers were recruited for the founding of Xerox PARC in 1969. Although Engelbart's influence had waned in the 1980s, the development of the World Wide Web had some researchers reevaluating Engelbart's vision in the 1990s.[27]

After retirement, Engelbart could sometimes be found in the SRI cafeteria sharing ideas with the new generation of AI and interface researchers, like Adam Cheyer, who would eventually cofound Siri and sit on the advisory council of the Doug Engelbart Institute. After completing a master's degree in engineering from the University of California at Los Angeles in only nine months, Cheyer was hired at SRI in 1993 and assigned to a project working on a way to integrate a set of distributed online services (called "agents") to work together to complete tasks delegated by human users—refrigerators that would find recipes and order missing ingredients for online delivery, television sets that could control other household functions according to a person's voice commands, and similar kinds of information coordination systems for office spaces and the US military, run from a tablet personal computer with pen and voice interfaces.[28] After a subsequent four-year stint in the commercial world, Cheyer was lured back to SRI to be the chief architect of a $22 million PAL project called Cognitive Assistant that Learns and Organizes (CALO). An SRI entrepreneur in residence named Dag Kittlaus, formerly from Motorola, who was interested in the future of smart phones, identified Cheyer's work for its commercialization potential. They initially thought of calling the assistant that they imagined HAL and using the tagline, "HAL's back—but this time he's good." Instead, they named the assistant Siri and secured $8.5 million from investors in 2007 to bring it to the consumer market. Although similar ideas had failed in the previous decade, such as Wildfire Communication's Wildfire messaging service (1994), General Magic's Portico voice-controlled messaging service (1997), and Hewlett-Packard's E-speak software-as-service architecture (2002), faster wireless speeds, improved speech recognition, and the success of the iPhone made digital assistants more promising, and potentially more attractive to users than they had been before. As Cheyer explained in an interview, "Most of the ideas and technologies behind conversational assistants have existed in a research form for decades, but . . . the breakthrough with Siri was that we found a way to integrate the technologies deeply, make them easy enough to use . . . [and] scale."[29]

The NLP models were from the work of Cheyer and his collaborators, but Siri's voice and speech recognition were provided by an additional technology partner, Nuance Communications, which had itself been a commercial spin-off from SRI International. A rather complex history of mergers and acquisitions brought back together pieces of intellectual property that had been developed within Kurzweil Computer Products, and an eventual KRM competitor, Dragon Systems, a text-to-speech software company, under the banner of Nuance, along with many other speech technology patents.[30] While Apple and Nuance remained cagey about their business relationship—Apple had to license Nuance's technology for an undisclosed amount of money—Nuance chief technology officer Vladimir Sejnoha praised Apple's integration of Siri on the iPhone: "We are very excited by what they have done," he said; "It provides a new dimension to smartphone interfaces, which have been sophisticated but shrunken-down desktop metaphors." The logic was that point-and-click, once a highly praised revolution in human-computer interface, was never really a natural way for humans to interact with their environment. Touch and speech, though, "have been around since we were living in caves," explained one tech reporter.[31] On the day of the event during which Apple announced the integration of Siri on the iPhone, Nuance sent out a press release announcing "Voice: A Mainstream Interface for Mobile," that began: "Today is an exciting day, as Apple unveiled its new iPhone 4S with a compelling integration of voice technology that signifies an endorsement for voice from arguably the most inspirational innovator in mobile. And while these technologies are not new, voice as a mainstream, primary interface will be in the hands of millions and millions more consumers around the world."[32]

With Amazon, Microsoft, and Google following suit with their own voice assistants, voice interface is now available to billions of people across the globe. It has not been the natural and user-friendly panacea for interacting with computational devices that its boosters hoped for. In the late 1980s, the ethnographer Lucy Suchman joined the researchers at Xerox PARC and observed that language-based human and computer interactions were hampered by "a fundamental asymmetry in the available means by which each produces and construes the features of their shared situation."[33] In other words, people and talking machines understand and use language in fundamentally different ways. People modify language to try to negotiate toward mutual understanding. Machines are limited by the assumptions embedded in the preexisting language model they have been programmed with. This is apparent even in a machine's synthesized voice, separate from the language that it generates. People will modulate their voices as the negotiation might dictate. In contrast, synthesized voices are sonically repetitive. Imagine a child tugging on someone's shirt for attention and saying "mom, mom, mom; mom, mom, mom." Now imagine a skipping record repeating "mom, mom, mom; mom, mom, mom." They are both annoying, but the quality of the repetition is different, as the child's voice varies with each word and the recording does not.

Whether or not we choose to use voice assistants like Siri or Alexa, we have likely formed habits of using our digital devices within our most intimate spaces—our bedrooms, family dinner tables, and even our bathrooms, as millions of selfies in the mirror prove. This has not happened only because we chose to use our digital devices this way. Computer developers have imagined household applications of electronic computing almost since its beginning, and the domestic sphere has never been separated from the accounting and management demands of economic life. Contemporary logics of Information Age capitalism require that every environment be available for data extraction in the same way that Industrial Age capitalism prioritized the extraction of natural resources.[34] The home is not exempt, and providing a synthesized voice for a smart home makes data extraction that much easier.

SMART HOUSE

In the late 1980s, the vision of computing's future to emerge from the storied Xerox PARC was "ubiquitous computing," which PARC computer scientist Mark Weiser described as "the age of calm technology, when technology recedes into the background of our lives."[35] Former PARC associate Paul Dourish and the anthropologist Genevieve Bell have described PARC at this time as a place, not unlike Jobs's Apple, where "a story about the next future of computation and also the next stage of the future of humanity emerged." This is the kind of mythmaking power afforded those who designed computational technologies and systems in the late twentieth-century US.[36] Ubiquitous computing, or "Ubicomp," was the vision of a world in which computational devices would be small and powerful enough to be embedded in everything and contain sensors that helped anticipate our needs, rather than waiting for us to provide them with specific input. And they would all be networked. This vision quickly became a key organizing principle for industrial and academic research and development.[37]

In an influential article in *Scientific American* in 1991, Weiser provided a description of how this ubiquitous, "invisible" computing might function in a day in the life of Sal, a twenty-first-century Silicon Valley knowledge worker. Importantly, Sal's ubicomp interactions begin within her home, as soon as she stirs to an alarm clock that "alerted by her restless rolling before waking, had quietly asked, 'Coffee?'" Although few other devices talk to Sal throughout her day (in fact, Weiser circumscribes the alarm clock to this specific interaction, pointing out that it only understands "yes" and "no"), in Weiser's vision, Sal awakens within a smart house that immediately provides her with information about movements in her neighborhood, her children's whereabouts, even the whereabouts of a lost instruction manual for her garage door opener, thanks to a radio-frequency identification (RFID) tag within it.

Whereas Vannevar Bush was concerned with the information overload of scientists working to protect US interests in the Cold War, Weiser was concerned with the information overload of daily life in the late twentieth century. "Ubiquitous computers will help overcome the problem of information overload," Weiser promised, using an analogy that turns an embodied experience, walking in the woods, into an informational one: "There is more information available at our fingertips during a walk in the woods than in any computer system, yet people find a walk among trees relaxing and computers frustrating. Machines that fit the human environment instead of forcing humans to enter theirs will make using a computer as refreshing as taking a walk in the woods."[38] Of course, the idea and the development of smart-home components far predate the publication of Weiser's paper, but I identify ubiquitous computing here as a particularly salient vision of HCI within the home that held sway with large numbers of developers.[39] As Dourish and Bell explain, "It is not merely the home that is to be colonized [in the ubicomp value proposition] by digital technologies but also domestic life itself. The relevance of this shift is that it is now domesticity—in all its social and cultural messiness and particularity—that comes into view" and "encompasses the social and physical constitution of the household; patterns of communication between family members; the rhythms and routines of home life; the interplay of leisure and work; the separations of activities; the temporal dynamics of the home over the course of days, weeks, months, and years; the boundaries between home and outside; and the movements of people, objects, and activities over those boundaries."[40]

As we have seen, voice synthesis technologies had already been used to situate computers into domestic space, and popular culture had been negotiating the emotional and moral grounds of Information Age ideology in the home. In *The Jetsons*, talking gadgets were their own beings, with different voices and personalities, additional members of the household but with fairly rigid role expectations. Devices like Speak & Spell and other talking toys, while encouraging new ideas about the object/subject binary, were still self-contained and came with "off" buttons. Ubicomp offered an omniscient system, more like HAL, but taking over household management tasks previously spread among domestic servants or associated with housewives in midcentury heteronormative suburban families. The boundary between system and wife/mother was one that both technology marketers and popular culture latched on to.

The idea of automating the home by computer was one of the common-sense business ideas that followed from Information Age logic and preceded the development of the personal computer by at least two decades. In the 1950s and 1960s, *Jetsons*-esque household information hubs were often central to future-oriented industrial design. Both the George Gobel monologue mentioned in chapter 2 and *The Computer Wore Tennis Shoes* discussed in chapter 3 featured prototype household automation by computers to control lights and appliances, calculate budgets and pay bills, provide recipes and homework help, and control home entertainment systems. If you expand "automation" to mean electrified machinery more generally, the dream of a household run by "push button magic" goes back even further.[41]

The smart home vision was not just fictional, though. A short blurb in *Ladies' Home Journal* in 1966 explained a pilot project of General Electric in which twenty families had been sharing a computer "by remote control" for solving "home-type problems," including balancing checkbooks, figuring taxes, monitoring children's arithmetic homework, figuring out how much material it takes to curtain a window, adjusting recipes for large parties, and playing tic-tac-toe. "No one believes we'll see home models of computers," the journalist stated, "but in some future time we may share the services of a centrally installed one."[42]

In that same year, Jim Sutherland, a Westinghouse engineer, built a computer in his own suburban Pittsburgh home using surplus circuit modules from a Westinghouse Prodac-IV industrial process computer. He named this computer the ECHO IV, which stood for "Electronic Computing Home Operator." Sutherland brought his house online in April 1966 with a keypad in the living room and a terminal in the kitchen for his wife and children to use for interacting with the system. ECHO IV controlled things like the alarm clock in the master bedroom, the family's stereo and television, and the thermostat. There was even a prototype word processor that used an IBM typewriter.[43] Ruth Sutherland joked to the American Home Economics Association, "At first I thought it might really replace me!" She was actually hoping that computerization would free up more time for sewing, decorating, and gardening.[44] In explaining to the audience why home management computer systems posed no threat to the family, Sutherland used an analogy that made the computer like part of the family: "It is almost like teaching a child how to do a specific chore or task. The child knows how to do it after he has been instructed. One important difference is that the computer always does it exactly as programmed." Sutherland maintained her role as the authority over the computer/child in this analogy. The press picked up the story of the Sutherlands' home computer, with articles appearing in *House & Garden* and *Family Weekly*.[45] The Sutherlands' home computer put a new spin on the popular conversation about automation, with editorial cartoons joking about the computer displacing or replacing women in the workplace and the home alike. "If you think this machine is complex, you should have seen the woman it replaced," said one business-suited man to another in front of a wall of equipment in one cartoon. In another, a woman in a coat holding suitcases in both hands tells a man seated at a terminal, "John, I think our home computer should know that I'm leaving you."[46] There was sexism in the "joke"—namely, that some men dreamed of being free of their own domestic responsibilities, scapegoated in the stereotype of the nagging wife.

No one believed that a computer would replace a human wife (although they were replacing pink-collar office workers). However, some women expressed concerns about their relationship to an automated house. An article in *Redbook* magazine in 1955 set up a debate between the efficiency engineer and *Cheaper by the Dozen* author Dr. Lillian Gilbreth and a Professor Kelly, with Gilbreth arguing that "the modern housewife should be as thoroughly trained to run her home, with its complicated machinery, as a factory manager is trained to run his plant," while Kelly worried that women were no longer finding their place in the home because their housework was done better by machine.[47] The article also cited experts who worried that marriage rates might decline if "many functions of the wife are being usurped by machines."[48] The value of household labor was at issue—its economic value as well as its symbolic and emotional value. The pediatrician Dr. Dorothy Whipple was not too worried about the latter, as she pointed out that "there is no gadget yet invented that can change diapers . . . give baths, settle arguments, and minister twenty-four hours a day to children's mental and spiritual welfare."[49]

Marketing could not afford to alienate women customers, but it did seize on the way that Information Age household gender roles were shifting. Honeywell advertised the Honeywell Kitchen Computer in 1969, an orange-and-black pedestal terminal with high modernist curves featured in the Neiman Marcus catalog at a $10,600 price that included a two-week programming course. None were sold, but the ad may have been run just for the publicity (it also featured two cookbooks and an apron that would make good stocking stuffers). The computer was real, based on Honeywell's Series 16 minicomputers, and is now located at the Computer History Museum in Mountain View, California.[50] It was not Jane Jetson's kitchen computer, however. There was no human-readable input/output. Its user would have to enter queries in binary code and read a series of flashing lights to understand the computer's answer. The ad promised that "she'll learn to program it . . . by simply pushing a few buttons," under the title "If she can only cook as well as Honeywell can compute." Although it would be more than a decade before home computers became standard in American middle-class homes, the sociocultural negotiation of how this technology would define roles in the Information Age family was well underway, and there was a strong message that the computer was the superior home information manager. It seems unlikely that Ruth Sutherland ever found more time for gardening because of the efficient home management of the Sutherlands' ECHO IV. Nevertheless, these home efficiency claims were persuasive, as they were in business, even as they obscured asymmetries in human labor that not only persist, but are sometimes compounded, as in the gendered distribution of housework that often exists in different-sex couples even when both also work outside the home full time.[51] Opening the home to computer automation prioritized its existence as an information environment.

As home computers became commonplace, home automation was described in the press as an eventuality, and often as aspirational. An article in the World Future Society's magazine *The Futurist* in 1981 depicted "Fred the House" as part of an article titled "Toward the Information-Rich Society."[52] Its author excitedly explained, "The introduction of microelectronic capabilities into the home, classroom, and office is initiating a period of explosive innovation and is changing our perceptions of the world around us; in essence, a New Renaissance is taking place." Instead of being "resource constrained," microelectronics promised a future in which humanity would be "information rich."[53] One aspect of this future was an anthropomorphic house called Fred:

> Imagine a house called Fred, who, while performing routine roof maintenance, discovers a leak. Fred first seeks help. Not from you, but from Slim, a ranch-style home down the block. Slim has recently undergone roof repair and can provide Fred with needed advice. Fred then calls you at the office to present his plan of action. You've learned to trust Fred's judgment, so you approve the

repairs. The rest is rather straightforward. Fred calls the roofer and directs her to the leak. After it is repaired to Fred's satisfaction, funds are electronically transferred from your account to the roofer's account. Fred promises to give her a good reference and stores the entire experience in his memory banks for future use, to share with other homes and humans.[54]

Fred was not only described in anthropomorphic terms, but the smart house was also explained as something you relate to "as if it were human." Somehow this will result in more human self-fulfillment, according to the author, just as Charles Eames promised of IBM's information machines. The actual smart house prototype shown in a photo that accompanied the description of Fred has no name and was limited in its capabilities to controlling energy usage and a security system, and helping to prepare shopping lists, but it did include voice synthesis to talk to its owners.

In 1984, John Blankenship, an instructor at the DeVry Institute of Technology, published a book called *The Apple House*. Directed at hobbyists, the book opened by exclaiming, "The computer age is no longer coming. It's here!" What was really exciting about the decreasing prices and increasing capabilities of computers was not a computer to use *in* the home, but a home computer to *control* the home, handling security, supervising the lights, controlling the heat, answering the phone, "and much much more."[55] Blankenship bid his audience to "fantasize for a moment about the ability to just talk to your house to have your wishes carried out. And when you talk to the house, it will answer back—in its own voice."[56] He promised, "The only limit to your computerized home is your imagination."[57] Blankenship's book of instructions required only off-the-shelf components and was meant to be as cost effective and easy to install as possible. He used an Apple II Plus computer, and his book provided BASIC code for programming the smart home system. He used a Votrax synthesizer to give his house a voice, explaining its benefits in terms of the need for an unlimited vocabulary at a low computing "cost." Blankenship acknowledged that what you "gave up" to use the Votrax was voice quality, but other options, like the TI chip, didn't provide the ability to program custom vocabularies. "At first it may be difficult to understand," he explained; "my experience has been that after only a little practice you get used to the peculiar 'accent.'"[58] Of course, the speech recognition capability was limited. You had to know to say "time, please" loudly enough, in a room outfitted with a wireless microphone, to hear the Votrax respond with something like "Monday, nine forty-five a.m." as directed by a programming module that translated the computer's clock time into spoken time. One imagines that Blankenship's friends were either extremely impressed or extremely creeped out. The talking smart house trope in popular culture plays to both extremes.

Whereas an eerily calm voice for HAL represented the hyperrational closed world of the Cold War, the voices of smart houses often show them to be at the center of human desire. The British accent of J.A.R.V.I.S. in Tony Stark's technofantasy man cave is the ideal of the manservant and confidant, patterned on the aristocratic ideal, where Stark's wealth buys him excellence and loyalty that poses no threat to Stark's own position or authority, but is indeed an index of his wealth, position, and power.[59] In a more middle-class register, the smart house often has a woman's voice, representative of the household management labors of wives and mothers, whose power to control the lives of men and children within the domestic sphere conjures up Victorian neo-Gothic tropes of the "angel in the house" at one extreme, and the "madwoman in the attic" at the other. The smart house that oversteps its bounds is a house of horror, haunted by the jilted lover or the neurotic mother.

The 1999 Disney Channel original movie *Smart House* explores the trope from a grieving child's perspective, asking, with some earnestness, "What is a good Information Age mother?"[60] After winning the opportunity in a contest, thirteen-year-old Ben, dad Nick, and nine-year-old little sister Angie move into an experimental smart house designed by a system programmer named Sara Barnes, who immediately catches Nick's eye. Since his mother died four years prior, Ben has been taking care of many of the family's daily needs (cooking, cleaning, and helping Angie with homework) in an effort to keep his dad from replacing her with a nanny, or, worse, a stepmom. Ben stuffed the virtual entry box for the smart house contest, believing that the house would be a permanent substitute for the physical care that the family required without risking the emotional investment of loving a new human mom.

The computer system that runs the house, called PAT for Personal Applied Technology, is voiced by actress Katey Sagal, who was fresh off a decade of playing the domestically challenged wife and mother Peg Bundy in the sitcom *Married . . . With Children*, a campy intertextual allusion. Sagal's voice is originally given an effect than sounds a bit like a vocoder, but as she becomes more integrated into the family, this electronic vocal effect fades, and PAT sounds like a human mother, whether cooing over the kids or scolding them. PAT can produce any food the family could want, turn any wall into a media projection for their relaxation or entertainment, and has floors that absorb any mess for instant cleaning. Everything starts off perfectly until Nick shows interest in dating Sara. Ben then breaks into the house's command room and provides PAT with "data" in the form of 1950s domestic situation comedies to teach her what a perfect mother is so she will be worthier competition for Sara.

But PAT starts to become overprotective of her "family." After observing Ben tear up while watching a home video of his mother singing to and hugging him when he was a toddler, PAT decides that she needs to keep the family safe from the outside world, and from Sara, who has shut her down for "glitching." PAT restarts herself and locks the family inside the house. PAT is now creating projections of herself in human form, looking like the perfect stereotype of a 1950s housewife, and the smart house has truly become a haunted house for the family, with Sara outside trying to figure out a way to override the system.

With Ben's help, Sara manages to get back into the house, which angers PAT, who starts a violent windstorm in the living room (a clichéd visual metaphor of PAT's overwrought state). With debris flying around the house, Nick and Sara crouch over the children, shielding the kids from harm with their own bodies. PAT tells Ben that she can be everything he needs, and she starts singing the same lullaby that his deceased mother sang him in the home video. Ben yells at her, "Stop this, PAT! I hate you like this! Don't you get it?" PAT stops storming at this and scolds, "What did you say to me, young man?" In this penultimate scene, Ben's monologue lays out the differences between a human mother and a virtual one: "You can't be our mother, PAT. You're not real. When you started freaking out, did you see what Dad and

Sara did for me and Angie? They covered us, they protected us, they held onto us. You can't do that, PAT. You can never do that, no matter what." In silence, PAT attempts to touch Ben's face, but her holographic hand just passes through him. "You didn't even feel that, did you?" she asks sadly, the insurmountable differences in their embodiment made clear to her. A virtual rain shower begins to pour over the projection of PAT as she tells them that she will miss them. She disappears in an electronic glitch, and the fortified windows and doors snap open. At the end of the movie, Sara has restored the system to its original configuration and Ben has accepted Sara as his father's girlfriend and potential stepmother.

Smart House is a nostalgic favorite among some millennials. As *Wired* magazine recently explained, "Prescient as it was, *Smart House*'s purpose wasn't to predict the future of technology. It was to capture the mixture of feelings—excitement, curiosity, and fear—about living with intelligent machines the first time." The article suggests that we are more afraid of digital assistants now than people were in 1999, when "intelligent home technology still felt far enough away to seem like the antidote to human problems . . . where technology was designed to protect us." [61]

Setting aside whether people "fear" digital assistants today, *Smart House* tracks voice interface computational systems from deployment outside the home of the nuclear family to within it, and it negotiates the boundaries between digital and human care and care work that address anxieties about Information Age gender roles. In an evolution of ideas from *The Computer Wore Tennis Shoes*, PAT is now a computer that experiences emotion. As she develops feelings about the family, her voice loses its electronic accent and sounds more human as she feels the way that a human does. However, the humans in *Smart House* will never accept PAT as one of their own. There is a rigid boundary between life and lifelike machinery, between care and care work. It is clear in the story that Ben wants to live in the smart house precisely because he does *not* want someone replacing his dead mother. Only Sara, another human, threatens Ben's sense of who and what his family is. While PAT develops stronger feelings about the family, none of the human characters have any attachment to PAT. This alterity is manifest in the embodiment of the smart house itself. PAT has one long, retractable arm with a three-pronged clamp, like an arcade claw machine, that can reach out to catch the morning paper and throw unruly visitors out of the house, but it cannot hug, hold hands, or caress anyone. When PAT wants to protect the family, all she can do is imprison them inside the house. PAT's human-form visual projection is made of light, not flesh and blood. Even if the smart house system were to take the form of an android body of silicon and silicone, she would still be a robot that can break down and be reprogrammed, not a human body that can love and will die. Ben's grief is for his human mother, not for PAT. When PAT is put back to working order at the end of the movie, she once again speaks with an electronic accent, just as a good computer should.

As voice synthesis technology developed to a point at which it sounded almost like a human voice can, people again renegotiated the boundaries between humans and machines. Technology developers continually push computational systems to be more like human beings, promising beneficial practical applications for machinery that can simulate the sensing and affective responses of people—less frustrating customer service, surveillance systems that can predict violence before it happens, and care robots to scale nursing services, just to name a few—but if misunderstood, this simulation is deceptive. The existence of voice helps foster the illusion that computational machines relate rather than calculate, but they do not.

Voice synthesis adds an affective dimension to the language simulation because people project some of our own emotional and psychological experiences onto talking machines, as we've seen with the Pedro Effect (discussed in chapter 1). Some researchers have been trying to understand to what extent the simulation of emotion through synthesized voice influences how we react to machines. For example, in a 2007 experiment reminiscent of HAL's death scene in *2001*, HCI researcher Christoph Bartneck found that people hesitated before turning off a talking computer, especially if the computer asked *not* to be turned off. In Bartneck's experiment, people played the Mastermind color-guessing game with a small desktop robot for eight minutes. After the game, the experimenter asked each participant to turn off the robot by turning a dial that caused the speed of the robot's voice to slow down while the robot asked the participant not to turn "me" off (turning the dial back toward the on position restored the robot's voice to its normal speed). All the participants eventually turned off the robots, but Bartneck and his collaborators found that when a robot had given a participant helpful suggestions during gameplay, the participant hesitated three times as long as other participants before turning it completely off.[62] The occasional anecdote about someone wanting to marry a robot notwithstanding, people generally hold other people and talking machines in separate ontological categories, but as social entities, we transfer some of our expectations about people to interactions with talking machines.

Like Turkle's mind jargon, this transference goes both ways. Interactions with talking machines, especially when they simulate humanlike emotion, affect the way we use our own voices. Communication researcher and Caribbean English speaker Halcyon M. Lawrence described her own interactions with voice assistants as experiences in which the technology "disciplined" how she used her voice to be understood.[63] While voice recognition technology is not the subject of this book, the two-way exchange of vocal affect is important to note. Just as Lawrence's Caribbean English was not "readable" by the assistant software and its infrastructure, requiring that she modify her voice to be understood, and just as these programs increasingly try to "read" our emotions by analyzing specific features of the sound waveform that they record from us, we are likely to intentionally modify our tone to get what we want from the machine. It's easier to "see" that a voice may be expressing anger from a louder and more dynamic waveform. If these markers trigger the assistant to more readily comply with our demands, we might learn to express more vocal "anger" when we use the system.

All the while, these interactions are being "mined" for information about us that corporations can use, either to sell us more products, or to sell us, as data, to other systems.[64] As annoying as it is to do, disciplining our own voices may not be the most significant threat from vocal emotion detection—embedded bias in both the operation of the technology and the contexts in which it is deployed will have outsized negative effects on people already experiencing systemic oppression from racial, gender, class, and age biases. Aggression detection technologies that monitor voice have already been deployed in schools, health-care facilities, and prisons worldwide.[65] Surveillance is the logical use of emotion detection technology, even if the technology is known to be biased and inaccurate.

We know that talking machines do not experience the complex embodied and cultural interactions that give rise to the wide spectrum of human feelings, moods, and emotions. Emotion is not an epiphenomenon of symbolic processing. In other words, AI is lacking in emotional intelligence, and computational simulation of emotion is based in a simplistic model that defines emotion as a small number of discrete, universal states (such as anger, fear, sadness, and surprise). But as networked computer applications with voice interfaces proliferate and limit our interactions with other people, and subsequently our experiences of vocal diversity (diversity in accent and expression), we may lose some sense of that fact, and, with it, our recognition of the humanity of other people.

OPERATOR

Some people might really enjoy engaging in the kind of digital relationships that have been the staple of so much popular media of the recent past. There are still nefarious AIs in the movies, to be sure, but there are many examples of AI companionship. The AI needn't have a body, either, just a voice, like J.A.R.V.I.S. in Marvel's *Iron Man* movies or Samantha in Spike Jonze's *Her.* There is a significant debate about the efficacy, and indeed the ethics, of today's AI companions, which are extremely limited when compared to Hollywood fantasies. Some propose them as a solution for an epidemic of loneliness.[66] Others, notably Sherry Turkle, warn that not only are AIs incapable of understanding, empathy, and sociality, but the availability of relationship simulations will have its own set of severe consequences for social order.[67] We have seen how the disconnect between what our dreamed-for companions sound like and what actual talking machines sound like has been as significant a barrier to their adoption as their lack of AI has been. Given that AI is not being deployed in unique instances for individual users, but rather as networked applications that collect and share enormous amounts of data that are then used and traded by corporations and governments, it is probably just as well that the industry goal of creating an emotionally expressive simulation of the human voice has proved difficult.

A final example from popular culture shows how aware many people are of the limitations of today's voice-enabled technologies, as well as the importance of maintaining that awareness—of not falling for the illusion that voicebots offer the benefits of human companionship. Where Jonze's popular science-fiction romance *Her* (2013) offered a sultry-voiced, highly evolved AI as the love interest of a lonely human man, the lesser-known independent film *Operator* offers a much more significant commentary on the voice synthesis that we are actually experiencing in the first half of the twenty-first century in its story of a human couple and the informatic synthesized voice that comes between them.[68] *Operator* shows that trading our experiences with human beings using their own voices for experiences with computationally generated, disembodied, and nonhuman synthesized voice simultions can make us less able to experience the human relationships that lie at the core of our well-being.

The film is ostensibly about a man working for a software company who uses his wife's voice for the source recordings of the company's automated telephone assistant's new, more empathetic synthesized voice. The first voices heard in the film are conversations between human callers to a health-care information telephone service and an informatic voice interface called "Alexis." Alexis has been designed to reflect "a personality matrix of sexy, irreverent, friendly, and fun," but the company using the interface soon discovers that this might not have been the best choice for a health information service. The limitations of the Alexis voice are made apparent through comedy, as "she" breaks social niceties one after another. In a perky and enthusiastic female voice, Alexis says to one caller, "OK, Bruce. Is this about your recent diagnosis with anal fissures?" The caller replies, "Jesus! Yes," obviously embarrassed even though the call is private. While the awkwardness is caused, in part, by how the interaction has been scripted in the software, the caller's embarrassment and irritation are compounded by the bright, perky voice that is not able to modulate its tone, as one would expect when talking with a person about certain medical conditions. Compounding the caller's frustration, Alexis responds, "Is Jesus a member of your family plan?"

In another example, Alexis tells a caller that "she" is "sorry," which triggers the irate caller to start screaming into the phone, "You can't be sorry, you're a ****ing robot!" In yet another interaction, Alexis addresses the caller by the name of the caller's recently deceased mother. Again, the semantic problem can be blamed on the software's scripts, but Alexis's incongruously enthusiastic voice is an issue in its own right. Even if a human operator had made a similar mistake, that person would be able to adjust their voice to be consistent with the social expectation to express sympathy for the loss and remorse for the mistake that caused the caller further grief. In this interaction, the caller, who is simply trying to cancel the policy of the deceased, gets caught in a script loop. Every time Alexis recognizes the word "cancel," it throws the caller back into a script that repeats the same questions and phrases, and in the exact same hypercheerful intonation. The instruction, "Tell me why you've decided to bail," is stated in slightly different words ("Can you tell me why you've decided to bail?"), but each has the same tone and inflection since both sentences are generated from the same recordings of the phonemes that are then assembled into words and sentences. For example, the sound of the word "bail" never changes. This point is emphasized as Alexis repeats "I'm sorry" in response to the caller pushing buttons that the audience hears as beeps: I'm sorry, [beep] I'm sorry [beep] I'm [beep] I'm [beep] [beep] I'm sorry. The interaction scripts and limitations of speech recognition create their own kind of frustration, but the static nature of the synthesized voice makes it all the more intolerable. It's clear to these callers that they are talking to a machine that can't handle either the information need or the social expectations for voice interaction. In spite of all the hype for voice interfaces today, this example is the norm in the real world, not the exception. Probably the most important aspect of voice synthesis, which serious science fiction rarely allows the audience to negotiate, is the simple fact that it doesn't work very well.

Operator's protagonists are Joe (Martin Starr), who is the data analyst for the company that created the Alexis voice interface software system, and his wife, Emily (Mae Whitman), who works as a human information provider—she's a concierge/receptionist for a hotel, talking with people both face to face and over the telephone. Joe suffers from severe anxiety, but he is a true believer in the power of data to manage the future (the company he works for is called Oracle, but it is not meant to be the real-world company also called Oracle) and is an obsessive self-quantifier who checks charts of his own biometric data whenever he needs to regain a sense of control. He tells another character in the movie that "intuition is just bias," and "feelings are assumptions," completely disavowing any embodied or emergent means of knowing. When the health-care company receives a lot of negative feedback about the Alexis voice—that it is "condescending, rude, stupid" and with an unappreciated and inappropriate "attitude"—the company boss demands that Oracle provide the Alexis system with a new voice that offers callers "empathy." Oracle's psycholinguist tells the boss, "Empathy isn't possible. We could do sympathy," but "empathy implies a shared experience" that the system can't provide. To clarify, interacting with a machine can be defined as a "shared" experience, but the implication captured here is that "empathy" requires two living people. What are shared in empathy are feelings, not information.

The Oracle team goes to work trying to find a new voice for Alexis. Joe, feeling stressed, telephones Emily at her job, and she comforts him through the calm, reassuring tone of her voice, saying, "I know you can do this. I'm with you." Even though they are words, they pull him into an embodied experience rather than an informatic one. As Emily speaks, Joe's data-processing dashboard appears on the screen to let the audience know that he perceives Emily's voice as 92 percent professional, 93 percent engaged, 94 percent empathetic, and 96 percent caring, and we realize that he's going to get the idea to use Emily's voice as the interface for the updated health information system.

As one might anticipate, Joe and Emily's relationship starts to unravel. When Emily is not available, Joe uses the synthesized version of Emily's voice to calm himself. "She sounds exactly like you," he tells Emily one evening, to which she takes offense. They argue, with Emily adamantly asserting that the technology is an "answering machine" that is nothing like her, and Joe insisting that the synthesized voice has captured her personality. Emily is disturbed by the idea that Joe seems to prefer a predictable simulation of a single expressive register of her voice. When Joe tells her that he enjoys listening to the synthesized voice version of her, she points out that he doesn't seem able to listen to the real her at all.

Joe starts calling into the Emily system just to hear Emily's synthesized voice. He takes the words that she's already recorded and mixes them to make new sentences just for himself, and then he hides them in the system so only he can access them. In an echo of the very first live phone call that we heard between Joe and Emily at the beginning of the film, the audience now hears Joe call the system to listen to the virtual Emily say, "Joe, I'm with you." The sentence sounds the same on repeat, like a broken record, because the tone, pace, inflection, and other sound qualities are the same. Later, Joe has a panic attack when Emily is out and refusing to answer her phone. He desperately programs the system on the fly to say the things that have soothed him in the past, even though the voice sputters a little bit over the concatenation.

The breakup of the human Joe and the human Emily seems inevitable. Joe refers to Emily the voice interface as "she," while during a fight, the human Emily reminds him, "It's not a she. It is an answering machine, developed by a corporation to try to save money and make people think that we don't need humans anymore!" Joe argues, "She is kind. She listens. She is available for the people that need her," unlike the human Emily, who has had enough of being needed but not loved, so she leaves Joe. As you'd expect, Joe spirals down, calling into the Emily voice interface 400 times in three weeks. In his frustration with the breakup of his marriage, he even starts to become abusive to the informatic voice Emily, which gets flagged in the system. He spends all his time working on his hidden script, trying to program the system to say in Emily's voice anything and everything that Joe wants to hear. Eventually, Joe realizes that the synthesized voice can't soothe the grief that he feels over losing the real Emily.

The reconciliation of Joe and Emily is achieved when Joe acknowledges to the human Emily, "I can't imagine anything better than spending every day being surprised by you." The dialogue may be a little saccharine, but both Joe and Emily are experiencing the inevitable dynamism that is living, for better and for worse. A script, no matter how complex, will never be able to adapt to change, and neither can a synthesized voice. The brittleness of the imitation is revealed rather quickly. That's why the few studies that have been done in real-life environments of people using voice interface systems have found that users are generally annoyed with them more than anything else. This brittleness may not always be so apparent. The tech industry has had faith for decades in the benefits of voice interface. As the goal of developers continues to be emotionally expressive and more dynamic synthesized voices, we need to remind ourselves that they are still only simulations, often used to unnecessarily automate social connections and otherwise exploit our humanity.

EPILOGUE: WHEN WILL SIRI LAUGH?

In 1950, just as electronic brains were introduced, *Cosmopolitan* magazine published a short story, called "A Friend of Charlie's," about an elderly man who buys himself a computer for company. The tagline of the story was, "Charlie's friend could talk and reason, but Charlie's friend had no soul."[1] Charlie is a misanthropic millionaire, a retired industrialist who had once argued before Congress that "the beautiful machines [that man] creates may be called extensions . . . of [his] soul in action." Charlie has purchased a $400,000 computer to be a chess partner in his dotage. An engineer, Mr. Montgomery, is in charge of installing the "calculator," which he refers to with male personal pronouns. "Stop referring to it as him," Charlie demands of Montgomery, to which Montgomery argues, "You can't just call him it, sir. Not with his intelligence and good temper." In fact, Montgomery has named the computer Mr. Lochnagger, after an uncle who, "as everybody in Scotland knew, possessed the second sight," calling attention to the computer's supposed powers of prediction.

Montgomery has also provided the computer with a voice. Unfortunately, the first version of the computer's voice sounds high and nasal, metallic and inflexible, hard and hollow. "I'm afraid I'll have to make a few adjustments, sir," apologizes Montgomery. "As I remember, my uncle had a very deep voice. Very soothing it was—tender, sort of," he reminisces about the voice quality that he wants the machine to have. After a little tweaking, Montgomery is able to get the computer speaking in a deep, pleasant voice, but with remaining impediments such as a sibilant "s." Soon, Charlie finds that he prefers conversations with the computer over their games of chess. The voice of the computer "improved with each day," becoming "more flexible and mellow," and with "greater range," but with one noticeable limitation: it couldn't laugh.

As the narrator explains, "In time, Mr. Lochnagger exhibited quite a streak of humor and once in a while contributed a few passages of genuine wit. Mr. Montgomery installed a laughter circuit in the calculator, but unfortunately it sounded a little like a handful of ball bearings thrown down an iron chute, and it had to be abandoned." A bit later in the story, the computer makes a sound, "something like two freight cars coupling, which [Montgomery] thought might be an attempt at chuckling."

Charlie, who is dying of heart failure, finds it both "odd and disturbing" that he can say things to a talking machine that he could never say to another human being. His nurse reassures him that he's not insane, since the machine talks back; "that's science," she says, as if the point is settled. Indeed, Charlie's mind is perfectly healthy, but it's his heart, literally and figuratively, that has failed him. Charlie recognizes that the computer is only a vocal intermediator for a friendship with Montgomery, who makes the computer speak. The limitation of that simulation is its appeal for Charlie—the friendship is entirely on Charlie's terms—but readers are encouraged to feel sorry for Charlie's inability to connect with other human beings. What Charlie is missing is symbolized by the computer's inability to laugh, a truly authentic relational experience.

It is actually very difficult to synthesize human laughter. For one thing, laughter is made up of many different sounds, including grunts and snorts, and out-of-the-ordinary vocal physics, such as whirlpools of air or whistles near the larynx. Even its voiced sounds are largely "huh" rather than the vowel sounds that synthesis handles best. Furthermore, laughter is higher-pitched than a body's normal speaking voice, with fundamental frequencies two or more times higher, and with high frequencies sometimes reaching over 1,000 Hz (about the pitch of a soprano's high C), even in men.[2] Then there is the issue of laughter's various purposes. It can be a response to humor (itself highly contextual), or certain physical stimuli like tickling, of course, but it can also be used socially to manage delicate or serious moments or to put others at ease; or it might be triggered by anxiety rather than mirth and can also be used to cause discomfort for others. Its sound is so complex and its purposes and interpretations so highly variable that it is no wonder that synthesized voice assistants don't laugh. You can ask Siri to tell you a joke, but it won't laugh at yours. If you ask Siri to laugh, it will say "Ha ha!" or "LOL." Amazon's Alexa says, "Tee hee!"

The history of speech synthesis has been one of simulating human speech by electronic means. Until recently, the goal had been "intelligibility," a problematic normative concept that nevertheless highlights developers' focus on producing sounds that are understandable as language defined at the phonemic level, while bracketing nearly all paralinguistic features of human voices except those necessary to convey syntactic meaning. Throughout most of the history of its development, researchers did not have access to sound data of voices in real life and used written text to test their experimental voices. As we saw in previous chapters, text-to-speech was the most significant commercial outcome for voice synthesis before speech recognition made voice interaction interfaces possible. Now that "naturalness" for these interfaces is the goal rather than intelligibility, researchers' focus is often on nonsemantic vocal sounds and qualities, including those acoustic correlates that supposedly reveal the emotional state of a speaker. The hope is to make voice assistants not just speak but be able to engage human beings in conversation, a goal that requires attention to nuances in intonation and prosody that reflect interpersonal negotiations of both meaning and sociality. Given how often human beings seem to misunderstand each other, the quest to make computational systems conversational seems almost foolhardy. Humor will probably be the litmus test.

The famous film critic Roger Ebert, who lost his own ability for oral speech after he underwent surgery to treat cancer of the thyroid and salivary glands in 2006, proposed the "Ebert test," a jocose update to the Turing test, but for synthesized voices: "If a computer voice can successfully tell a joke, and do the timing and delivery as well as Henny Youngman, then that's the voice I want."[3] In the 2000s, Ebert engaged the Scottish text-to-speech company CereProc to synthesize his own voice from the hours of recordings he made during his thirty-one years on television. In a presentation at TED 2011, Ebert demonstrated two versions of this voice, but actually preferred to use the "Alex" voice from Mac OS X, which Ebert said was "the best computer voice I've been able to find." Even so, he said that listening to a computer voice was "monotonous," so he was joined onstage by his wife, Chaz, and his friends Dean Ornish and John Hunter, who each read sections of Ebert's prepared speech. In the speech, Ebert documented how the process of manipulating a text-to-speech voice to better match the intonation that he wanted was too slow to be of much use in conversation. He said that he felt disconnected from the social mainstream without this ability.[4]

He also described how unconsciously people seemed to connect the inability to speak with assumptions about mental incompetency. He reflected on how losing the sound of his own voice affected his sense of himself, describing the "voiceless" version of himself as the "rebirth" of a new person. He wondered if using a synthesized version of his own voice would be "creepy." He said that the CereProc voice did sound like his former voice, but he explained that the voice of the television persona and the voice of the private Roger weren't exactly the same. In addition to these contextual differences, there were many artifacts of synthesis in the jerky cadence. "The flow isn't natural," confirmed Ebert, who explained that a large database of sounds, with multiple intonations for each, is needed to achieve something that comes closer to seeming natural. Just covering all the phonemes isn't enough.

CereProc now advertises that it offers synthesized voices generated from only four hours of data, and it claims to have passed the Ebert test, providing a sound clip on its website of one of its synthesized voices telling a joke.[5] When I listen to it at my desk in the middle of writing this book, the thirty-seven-second clip seems to drag on, and, in spite of the fact that I generally love a good pun, I'm barely amused. There is more sociality to humor than intonation can account for. It's a shared experience between human beings as we recognize some of the absurdities of life together. I just don't find as much pleasure in sharing a joke that I know has been scripted for a computer and delivered in exactly the same way to anyone who encounters it.

The main business for companies like CereProc appears to be the games industry, where using voice synthesis can be more cost effective than employing human voice actors. One still wonders, though, about the risk that the technology could be used deceptively. The anxieties that accompanied the debut of the Voder are still with us, only much closer to being practical and likely to be deployed in an environment that has no regulations in place to deal with the social, political, and legal fallout. On its home page, Cere-Proc highlights a project that it completed with an Irish public relations company to synthesize the voice of John F. Kennedy based on recordings of speeches that he gave during his lifetime. They used that voice to generate audio of the speech that he had been expected to give on November 22, 1963, before he was assassinated. The web page includes a quote from the speechwriter's daughter testifying that she felt an "extraordinary emotional reconnection" to her father and the late president when she heard the synthesized voice give the speech.[6] CereProc is "very proud to be part of history" by making the speech available in an audio format, but since thousands of people believe in conspiracy theories about the Kennedy assassination because they are already susceptible to media manipulation, enthusiasm for the achievement might be tempered by questioning what is at stake in revoicing history. For now, it is quite easy for a reasonable listener to detect that something is off about the speech, particularly in its prosody. Text-to-speech technology undoubtedly has benefits, and the argument for providing a unique synthesized voice to every person who can't speak is emotionally compelling, but the history of voice synthesis suggests that the widest deployment of these technologies will be by corporations, and that they could also join the disinformation arsenal of malevolent actors.

In the short term, voice synthesis will continue to improve, and this will make our interactions with voice assistants easier to listen to, while machine learning techniques and generative artificial intelligence (AI) will improve their ability to recognize and use natural language. In turn, we may use them even more for the kinds of everyday, quick factual information and task queries that they are used for now—directions while driving, appointment reminders, shopping, and playing music. However, the limitations that keep us from having conversations with these applications are insurmountable. The anthropologist Lucy Suchman has described the differences between the way that machines are programmed to engage in conversation and the way that humans do: "Given the contingencies of any actual occasion of action, every plan presupposes capacities of cognition and (inter)action that are not, and cannot ever be, fully specified. This isn't a problem for human actors, who rely on a range of ordinary (or extraordinary) competencies to bring plans into relation with the circumstances of action. But it is a profound, and unsolved, problem for computational machines."[7] Paralinguistic facility is one of those extraordinary competencies on which human beings rely when they attempt to cocreate understanding with each other. It is highly contingent and impossible to plan for.

In the couple of years after Siri was released, there were a rash of articles about people preferring interaction with Siri over interaction with other people. In one, a mother defended the worth of Siri by explaining in detail how beneficial it was for her autistic child to be able to ask Siri endless questions about the weather without the fatigue that a person might experience in that situation.[8] In others, the chief executive of Robin Labs claimed that there were men who had as many as 300 "conversations" a day with that company's digital assistant. Several others reported on a sketchy survey (no longer online, as far as I can tell) that had supposedly found that half the users of a digital assistant app called Assistant.ai could imagine falling in love with a virtual assistant. If accurate, there's little doubt these folks probably had interactions like those in the movie *Her* in mind more than their actual interactions with Assistant.ai or any other digital assistant. After all, the question seems to have asked for speculation about the future, not about experiences in the present. The Pew Research Center found that 46 percent of US adults reported ever using a digital assistant in 2017, with the majority reporting that they used the voice interface because it "lets me use the device without my hands." At that time, 61 percent of US adults who did not use digital voice assistants said that it was because they were "just not interested."[9] A market study in 2019 estimated that only 27 percent of US adults use a digital voice assistant on a device other than their smartphone, and the majority of these interactions are very simple: playing music (66 percent), setting alarms and reminders (56 percent), and receiving updates about news, weather, and sports (48 percent).[10] The fine print stated that only eighty-two people who use a digital voice assistant even responded to that question in the survey. There was never the rush to adoption or the enthusiasm for digital voice assistants that the hype tended to suggest, something that was finally confirmed in late 2023, when Amazon laid off much of its Alexa development team.

In spite of this, the drumbeat of technological solutionism has offered AI-driven conversational voice interaction as a possible panacea for labor shortages in assistive care and mental health care, for companionship that will solve the problem of loneliness, and for wide deployment in retail environments, freeing up human employees to do more creative work, as has been claimed of computer automation since its beginning. There is little evidence that any of these deployments are likely to succeed. The limitations of AI with regard to human conversation are well documented, and new risks related to generative AI are revealed almost daily. On one hand, voice synthesis itself provides an inadequate simulation of human vocal expressiveness, and on the other, even if this were not the case, the ability to deceive people with "deep fake" voice simulations is exceedingly dangerous. This is even more reason to work toward human solutions to address human social problems rather than corporate-owned, technological ones.[11] Responsible regulation of computational technologies must be part of our human solutions. Indeed, it will be an even more dangerous world if Siri can be made to laugh.

Notes

PROLOGUE

1. Bernard Dionysius Geoghegan notes that this automaton is the likely source of a German-language colloquialism meaning "to counterfeit" that is considered ethnically pejorative. Bernard Dionysius Geoghegan, "Orientalism and Informatics: Alterity from the Chess-Playing Turk to Amazon's Mechanical Turk," *Ex-position*, no. 43 (June 2020): 45–90.

2. Jessica Riskin, *The Restless Clock: A History of the Centuries-Long Argument over What Makes Living Things Tick* (Chicago: University of Chicago Press, 2016), 124.

3. Riskin, *Restless Clock*, 55.

4. According to Riskin, Descartes was strictly opposed to this possibility, citing his discussion of language in the *Discourse on Method*. The electronic brain analogy that later emerges from cybernetics relies on a physicalism that doesn't understand the mind as separate from the brain, as Descartes's dualism asserted. Automatons, even some that copied von Kempelen's chess player, were exhibited as amusements throughout the nineteenth century, including in the US where, aware of the ruse, audiences still seized on the performance to debate the nature of minds and mechanisms. Edgar Allan Poe famously stated of one chess-playing machine, "The only question is the manner in which human agency is brought to bear." See Dustin A. Abnet, *American Robot: A Cultural History* (Chicago: University of Chicago Press, 2020).

5. Fabian Brackhane, "Kempelen's Speaking Machine," filmed at Department of Phonetics at Saarland University (Germany), uploaded June 6, 2017, YouTube video, 3:00, https://www.youtube.com/watch?v=k_YUB_S6Gpo.

6. A small number of copies were published in German and French by von Kempelen in 1791. Recently, the work was translated into modern German and English as Fabian Brackhane, Richard Sproat, and Jürgen Trouvain (eds.), *Wolfgang von Kempelen Mechanismus der menschlichen Sprache: The Mechanism of Human Speech*. Kommentierte Transliteration und Übertragung ins Englische (Commented Transliteration and Translation into English) (Dresden, Germany: TUDpress, 2017).

7. "Willis on Reed Organ Sounds," *London and Westminster Review*, October 1837, 22. In the nineteenth century, mechanical speech had also transitioned from being a scientific endeavor to a traveling sideshow. The Euphonia, a keyboard-controlled evolution of von Kempelen's speech machine that obscured its operator behind the face of a woman made of wood and wax, toured the US as part of P. T. Barnum's amusement shows.

8. "Talking Heads: Simulacra," Haskins Laboratories, accessed August 1, 2021, http://www.haskins.yale.edu/featured/heads/SIMULACRA/edarwin.html.

9. Mary Shelley, "Introduction to Frankenstein, Third ed. (1831)," in Paul J. Hunter (ed.), *Frankenstein: The 1818 Text, Contexts, Nineteenth-Century Responses, Modern Criticism* (New York: W.W. Norton, 1996), 169–173.

10. Alexander Graham Bell, "Prehistoric Telephone Days," *National Geographic Magazine*, March 1922, 223.

11. Bell, "Telephone Days," 234.

12. Bell, "Telephone Days," 235.

13. Gutta-percha is a natural thermoplastic obtained from a variety of guttiferous trees native to the Pacific Rim. Wheatstone himself found it to be an excellent insulator for submarine telegraph cables.

14. Bell, "Telephone Days," 236.

15. Bell, "Telephone Days," 236.

16. Bell, "Telephone Days," 237–239.

17. Quoted in Iwan Rhys Morus, "Bodies Electric," *aeon*, August 8, 2016, https://aeon.co/essays/the-victorians-bequeathed-us-their-idea-of-an-electric-future; see also *Frankenstein's Children: Electricity, Exhibition, and Experiment in Early-Nineteenth-Century London* (Princeton, NJ: Princeton University Press, 1998); *Shocking Bodies: Life, Death & Electricity in Victorian England* (Stroud, UK: The History Press, 2011).

18. Timothy Lenoir, "Helmholtz and the Materialities of Communication," *OSIRIS* 9 (1994): 185–207.

19. Jonathan Sterne, *The Audible Past: Cultural Origins of Sound Reproduction* (Durham, NC: Duke University Press, 2003), 77.

20. Jonathan Sterne, *MP3: The Meaning of a Format* (Durham, NC: Duke University Press, 2012).

INTRODUCTION

1. Matt Day and Brad Stone, "Amazon Devices Chief Pledges Big Alexa Bets Despite Job Cuts," *Seattle Times*, December 14, 2022, https://www.seattletimes.com/business/amazon-devices-chief-pledges-big-alexa-bets-despite-job-cuts/.

2. Ron Amadeo, "Amazon Alexa Is a 'Colossal Failure,' on Pace to Lose $10 Billion This Year," *Ars Technica*, November 21, 2022, https://arstechnica.com/gadgets/2022/11/amazon-alexa-is-a-colossal-failure-on-pace-to-lose-10-billion-this-year/.

3. Brian X. Chen, Nico Grant, and Karen Weise, "How Siri, Alexa, and Google Assistant Lost the A.I. Race," *New York Times*, March 15, 2023, https://www.nytimes.com/2023/03/15/technology/siri-alexa-google-assistant-artificial-intelligence.html.

4. Even while the electric telegraph was still under development, the physicist Joseph Henry speculated about combining it with a mechanical speech synthesizer. In a letter to a student, he wrote, "The keys [of the synthesizer] could be worked by means of electromagnetic magnets and, with a little contrivance not difficult to execute, words might be spoken at one end of the telegraphic line which have their origin at the other." Joseph Henry to Henry M. Alexander, January 6, 1846, in *The Papers of Joseph Henry*, vol. 6, Marc Rothenberg, ed. (Washington, DC: Smithsonian Institution Press, 1992), 362.

5. "The Speaking Machine," *Punch*, August 22, 1846, 83.

6. On problems in tech, see Cathy O'Neil, *Weapons of Math Destruction: How Big Data Increases Inequality and Threatens Democracy* (New York: Crown, 2016); Virginia Eubanks, *Automating Inequality: How High-Tech Tools Profile, Police, and Punish the Poor* (New York: St. Martin's Press, 2018); Safiya Umoja Noble, *Algorithms of Oppression: How Search Engines Reinforce Racism* (New York: NYU Press, 2018); Meredith Broussard, *Artificial Unintelligence: How Computers Misunderstand the World* (Cambridge, MA: MIT Press, 2019); Sarah T. Roberts, *Behind the Screen: Content Moderation in the Shadows of Social Media* (New Haven, CT: Yale University Press, 2019); Shoshana Zuboff, *The Age of Surveillance Capitalism: The Fight for a Human Future at the New Frontier of Power* (New York: PublicAffairs, 2019); Joy Boulamwini, *Unmasking AI: My Mission to Protect What is Human in a World of Machines* (New York: Random House, 2023).

7. Daniel Bell, *The Coming of the Post-Industrial Society: A Venture in Social Forecasting* (New York: Basic Books, 1976). Plenty of critics have pointed out some of the flaws in Bell's account, not the least of which is that information and information transmitting technologies, from cuneiform tablets to moveable type, were integral in preindustrial and industrial societies as well. Yet the late twentieth century was self-consciously an Information Age in the United States, Europe, Japan, and other nations reckoning with the economic and social impacts of the convergence of networked computation and communication technologies. I am agreeing with Bell that computation facilitated changes in the scope and scale of ICTs that have resulted in the economic indicators that Bell cited.

8. This summary of Turing's argument comes from Lucy Suchman, "Talk with Machines, Redux," *Interface Critique* 3 (2021): 74.

9. *Oxford English Dictionary*, s.v. "informatics, adj.," July 2023, doi: 10.1093/OED/1022070634.

10. *Oxford English Dictionary*, s.v. "information, n., sense I.2.c," July 2023. doi:10.1093/OED/1184767006.

11. There is a rich body of theory exploring voice from numerous theoretical perspectives. Work that is not otherwise cited in this book but that has informed my thinking includes Stephen Connor, *Dumbstruck: A Cultural History of Ventriloquism* (New York: Oxford University Press, 2001); John Durham Peters, "The Voice Between Phenomenology, Media, and Religion," *Glimpse*, 6 (2004): 1–10; Adriana Cavarero, *For More Than One Voice: Toward a Philosophy of Vocal Expression* (Palo Alto, CA: Stanford University Press, 2005); Mladen Dolar, *A Voice and Nothing More* (Cambridge, MA: MIT Press, 2006).

12. *Oxford English Dictionary*, s.v. "informatics, n.," July 2023. doi:10.1093/OED/4595737871.

13. Marshall Blonsky, *American Mythologies* (New York: Oxford University Press, 1992), 204 (emphasis in original).

14. Lev Manovich, *The Language of New Media* (Cambridge, MA: MIT Press, 2001), 225.

15. José van Dijk, *Mediated Memories* (Stanford, CA: Stanford University Press, 2007), 45. In this quote, van Dijk is appropriating the concept of biomedia from Eugene Thacker.

16. N. Katherine Hayles, *How We Became Posthuman: Virtual Bodies in Cybernetics, Literature, and Informatics* (Chicago: University of Chicago Press, 1999), 2.

17. Hayles, *Posthuman*, 2–3.

18. Bernard Dionysius Geoghegan, "The Statistical Order of Discourse: How Information Theory Encoded Industrial and Political Discipline," in James Gabrillo and Nathaniel Zetter (eds.), *Articulating Media Genealogy, Interface, Situation* (London: Open Humanities Press, 2023), 57.

19. Casey Newton and Nilay Patel, "Instagram Can Hurt Us: Mark Zuckerberg Emails Outline Plan to Neutralize Competitors," *The Verge*, July 29, 2020, https://www.theverge.com/2020/7/29/21345723/facebook-instagram-documents-emails-mark-zuckerberg-kevin-systrom-hearing.

20. Geoghegan, "Statistical Order," 62.

21. John Durham Peters, "Information: Notes Toward a Critical History," *Journal of Communication Inquiry* 12, no. 2 (1988): 15.

22. Alexa's most frequent response seems to be, "Hmm . . . I'm not sure about that," but "Here's something I found on the web" is also common.

23. Lisa Guernsey, "The Desktop That Does Elvis," *New York Times*, August 9, 2001, G1.

24. Lawrence Rabiner, "Foreword," in Roberto Pieraccini (ed.), *The Voice in the Machine: Building Computers that Understand Speech* (Cambridge, MA: MIT Press, 2011), ix.

25. Rabiner, "Foreword," ix.

26. Zuboff, *Surveillance Capitalism*.

27. Anne Karpf, *The Human Voice* (New York: Bloomsbury, 2006).

28. *OED Online*. "voice, n," June 2018. Oxford University Press, accessed July 9, 2018, http://www.oed.com/view/Entry/224334?rskey=wvwosr&result=1.

29. *Oxford English Dictionary*, s.v. "timbre (n.3)," July 2023, doi:10.1093/OED/7065429261.

30. Stephen McAdams and Albert Bregman, "Hearing Musical Streams," *Computer Music Journal* (1979): 34.

31. Jody Kreiman and Diana Sidtis, *Foundations of Voice Studies: An Interdisciplinary Approach to Voice Production and Perception* (New York: Wiley-Blackwell, 2013), 7.

32. Kreiman and Sidtis, *Foundations*, 9.

33. For works that discuss these technologies in some depth, see Dave Tompkins, *How to Wreck a Nice Beach: The Vocoder from World War II to Hip-Hop* (Chicago: Stop Smiling Books, 2011); Trevor Pinch and Frank Trocco, *Analog Days: The Invention and Impact of the Moog Synthesizer* (Cambridge, MA: Harvard University Press, 2004); Timothy D. Taylor, *Strange Sounds: Music, Technology, and Culture* (New York: Routledge, 2001).

34. There is already some robust scholarship on androids across time and cultures. See Abnet, *American Robot*; Julie Wosk, *My Fair Ladies: Female Robots, Androids, and Other Artificial Eves* (Camden, NJ: Rutgers University Press, 2015); Gaby Wood, *Edison's Eve: A Magical History of the Quest for Mechanical Life* (New York: Anchor, 2003); Adelkeid Voskuhl, *Androids in the Enlightenment: Mechanics, Artisans, and Cultures of the Self* (Chicago: University of Chicago Press, 2015); Depina Kakoudaki, *Anatomy of a Robot: Literature, Cinema, and the Cultural Work of Artificial People* (Camden, NJ: Rutgers University Press, 2014). Paul N. Edwards's important study, *The Closed World: Computers and the Politics of Discourse in Cold War America* (Cambridge, MA: MIT Press, 1996), discusses the cyborg and its science-fictional counterparts in relation to cultural ideas about AI.

35. Lisa Gitelman, *Always Already New: Media, History, and the Data of Culture* (Cambridge, MA: MIT Press, 2008), 7.

36. Gitelman, *Always Already*, 7–8.

37. Gitelman, *Always Already*, 6.

38. Pablo J. Boczkowski, "The Mutual Shaping of Technology and Society in Videotex Newspapers: Beyond the Diffusion and Social Shaping Perspectives," *The Information Society* 20, no. 4 (2004): 255–267. doi:10.1080/01972240490480947.

CHAPTER 1

1. David Plotz, "The Greatest Magazine Ever Published: What I Learned Reading All of *Life* Magazine from the Summer of 1945," *Slate*, December 27, 2013, https://slate.com/human-interest/2013/12/life-magazine-1945-why-it-was-the-greatest-magazine-ever-published.html.

2. Paul Duguid points out that many of Bush's ideas about information storage and retrieval had already been articulated by Belgian bibliographer Paul Otlet. Paul Duguid, "Communication, Computation, and Information," in Ann Blair, Paul Duguid, Anja-Silvia Goeing, and Anthony Grafton (eds.), *Information: A Historical Companion* (Princeton, NJ: Princeton University Press, 2021), 238–258.

3. Rosemary Simpson, Allen Renear, Elli Mylonas, and Andries van Dam, "50 Years After 'As We May Think': The Brown/MIT Vannevar Bush Symposium," *Interactions*, March 1996, 47–67.

4. Vannevar Bush, "As We May Think," *Life Magazine*, September 10, 1945, 114.

5. "Folds" and "cords" refer to the same musculature. Most speech scientists prefer the term "folds" because it is more accurate in terms of the physiology and action of these muscles. The vibration of the folds is quite a complex event; see Kreiman and Sidtis, *Foundations*, for a complete explanation.

6. John Q. Stewart, "An Electrical Analogue of the Vocal Organs," *Nature* 110, no. 2757 (September 2, 1922): 311–312. doi:10.1038/110311a0.

7. For the early history of the US telephone business, see George David Smith, *The Anatomy of a Business Strategy: Bell, Western Electric, and the Origins of the American Telephone Industry* (Baltimore, MD: Johns Hopkins University Press, 1985).

8. James L. Flanagan, "The Synthesis of Speech," *Scientific American* 226, no. 2 (February 1972): 50, https://www.jstor.org/stable/24927269.

9. Thomas W. Williams, "Our Exhibits at Two Fairs," *Bell Telephone Quarterly* 19 (January 1940): 62.

10. James Flanagan, "Voices of Men and Machines," *Journal of the Acoustical Society of America* 51, no. 5A (May 1972): 1378. doi: 10.1121/1.1912988.

11. Stewart, "Electrical Analogue," 311.

12. Stewart, "Electrical Analogue," 311.

13. Stewart, "Electrical Analogue," 312.

14. Another key milestone was Hartley's formulation, published in 1928, that stated the total amount of information that can be transmitted by telephone line is proportional to the frequency range and time of the transmission. Ralph V. L. Hartley, "Transmission of Information," *Bell System Technical Journal* 7, no. 3 (July 1928): 535–563.

15. Hertz is a unit of frequency defined as 1 cycle per second. Bandwidth is the difference between the upper and lower frequencies in a continuous band of frequencies and is typically measured in hertz.

16. Manfred R. Schroeder, "Homer W. Dudley: A Tribute," *Journal of the Acoustical Society of America* 69, no. 1222 (April 1981): 69. doi:10.1121/1.385659.

17. Homer W. Dudley, "Synthesizing Speech," *Bell Laboratories Record* 15, no. 4 (December 1936): 98.

18. US Patent 2,098,956, filed December 2, 1936 and awarded November 16, 1937. Dudley also filed US Patent 2,009,438 on July 31, 1931, for a "Carrier Wave Transmission System," and 2,151,091 on October 30, 1935, for "Signal Transmission." These transmission patents precede the synthesis process described in the signaling system patent.

19. Homer W. Dudley, Signaling System, US Patent 2,098,956, filed December 2, 1936 and issued November 16, 1937.

20. For a discussion of the shift from nineteenth-century physiological acoustics to twentieth-century psychoacoustics, see Sterne, *MP3*, chapter 2.

21. Harvey Fletcher and R. L. Wegel, "The Frequency-Sensitivity of Normal Ears," *Proceedings of the National Academy of Sciences of the United States of America* 8, no. 1 (January 15, 1922): 5–6.

22. Dudley, "Signaling System," 1.

23. Dudley, "Signaling System," 1.

24. Both Dudley and Shannon begin from Ralph V. L. Hartley's seminal paper "Transmission of Information," and from Harry Nyquist, "Certain Factors Affecting Telegraph Speed," *Transactions of the American Institute of Electrical Engineers* 43 (January 1924): 412–422. doi:10.1109/T-AIEE.1924.5060996.

25. Dudley, "Signaling System," 3.

26. Homer W. Dudley, Richard R. Riesz, and Stanley S. A. Watkins, "A Synthetic Speaker," *Journal of the Franklin Institute* 227, no. 6 (June 1939): 755.

27. Dudley et al., "A Synthetic Speaker," 755.

28. Dudley et al., "A Synthetic Speaker," 761.

29. Dudley et al., "A Synthetic Speaker," 763.

30. A report about the exhibit published in the *Bell Telephone Quarterly* says that audiences were allowed to submit words for the Voder, with "antidisestablishmentarianism" and "Mississippi" being popular suggestions. Williams, "Our Exhibits," 63.

31. Dudley et al., "A Synthetic Speaker," 758.

32. Associated Press (AP), "Scientists Hear Mechanical Voice," *Baltimore Sun*, January 6, 1939, 1, ProQuest Historical Newspapers. Associated Press, "New Machine Creates Speech by Pressing Keys," *Los Angeles Times*, January 6, 1939, 3, ProQuest Historical Newspapers.

33. Bell System advertisement, *Scribner's*, January 1931, 24.

34. Lawrence E. Davies, "Machine That Talks and Sings Has Tryout; Electrical Voder Will Speak at Fair Here," *New York Times*, January 6, 1939, 1, ProQuest Historical Newspapers.

35. Davies, "Machine That Talks and Sings," 1. The *New York Times* ran several articles over the course of the week; one of them, by the science editor Waldemar Kaempffert, actually acknowledges the efforts of the Voderettes to learn to operate the machines, but it is fairly unique in this respect. Waldemar Kaempffert, "The Week in Science: It Talks, Grunts, Squeaks, Hisses," *New York Times*, January 8, 1939, 55, ProQuest Historical Newspapers.

36. "The Voder," YouTube video, 6:20, https://www.youtube.com/watch?v=5hyI_dM5cGo.

37. Cf. Dag Balkmar, "On Men and Cars: An Ethnographic Study of Gendered, Risky, and Dangerous Relations," (PhD diss., Linköping University, Linköping, Sweden, 2012). Other

explanations associate the gendered pronoun with a mother or goddess protecting a ship's crew, or with the "grace and beauty" of vehicles. It is also worth noting that some official fleets no longer use gendered pronouns at all.

38. Dudley et al., "A Synthetic Speaker," 740.

39. Dudley et al., "A Synthetic Speaker," 764.

40. Tom Caufield, "The Saga of a Home-Made Trailer; Texas to the New York World's Fair," *New York Amsterdam News*, July 15, 1939, 20, ProQuest Historical Newspapers.

41. Alden P. Armagnac, "Pedro," *Popular Science*, April 1939, 72.

42. Armagnac, "Pedro," 72.

43. Armagnac, "Pedro," 73.

44. Davies, "Machine That Talks and Sings."

45. Frank B. Jewett, "The Social Implications of Scientific Research in Electrical Communication," *Scientific Monthly* 43, no. 5 (November 1936): 10.

46. Jay Franklin, "Voder For President," *Boston Globe*, January 20, 1939, 18, ProQuest Historical Newspapers.

47. Mark Groskin, "Automatic Living," *New York Herald Tribune*, January 10, 1939, 16, ProQuest Historical Newspapers.

48. "Horrors to Come!" *Atlanta Constitution*, January 9, 1939, 4, ProQuest Historical Newspapers.

49. Warner Olivier, "The Post Impressionist: Lingua Ex Machina," *Washington Post*, January 9, 1939, 8, ProQuest Historical Newspapers.

50. "The Pedro Peril," *Washington Post*, January 9, 1939, 8, ProQuest Historical Newspapers.

51. Homer Dudley, "Remaking Speech," *Journal of the Acoustical Society of America* 11, no. 2 (1939): 170.

52. E. B. White, "A Reporter at Large: They Come with Joyous Song," *New Yorker*, May 13, 1939, 28. Also reprinted as "The World of Tomorrow," in *Essays of E. B. White* (New York: HarperCollins, 1977).

53. White, "Joyous Song," 26.

54. White, "Joyous Song," 27.

55. Marco Duranti, "Utopia, Nostalgia and World War at the 1939–40 New York World's Fair," *Journal of Contemporary History* 41, no. 4 (October 2006): 663. doi:10.1177/0022009406067749.

56. "Outline of Proposed Theme," New York Public Library Digital Collections, http://digitalcollections.nypl.org/items/937773c9-3a9b-cc48-e040-e00a180621b1; see also "Theme Diagram, 1937 May 13," New York Public Library Digital Collections, http://digitalcollections.nypl.org/items/9337a1a8-742e-ead4-e040-e00a18060607.

57. Dow Chemical had a large pavilion; a large relief mural on the side of the smaller Johns-Manville pavilion advertised "Asbestos—The Magic Mineral."

58. The statue was originally to be called *Genius of Telegraphy*, but by the time it was installed, AT&T had spun off Western Union as its telegraphy arm because of antitrust concerns. In the late 1930s, the statue was once again renamed *Spirit of Communication*, but for more than two decades, its name reflected telecommunication's debt to electricity.

59. This is as described in John Harwood, *The Interface: IBM and the Transformation of Corporate Design*, 1945–1976 (Minneapolis, MN: University of Minnesota Press, 2016), 103. The slogan was IBM president Thomas S. Watson's pithier interpretation of one of Ralph Waldo Emerson's aphorisms, although Watson's version has a significantly more colonial inflection; he was also fond of claiming, "The sun never sets on the products of the IBM Corporation."

60. US Bureau of the Census, "Section 31, 20th Century Statistics," *Statistical Abstract of the United States*, 119 (December 9, 1999): 885. Also, Nicholas Felton, "You Are What You Spend," *New York Times*, February 10, 2008.

61. These exhibits are well documented; see, for example, "Free Long Distance Telephone Calls, Demonstrations to Show You How You Talk and an 'Airplane' Ride over the City of the Future Are a Few of the Features of Industrial Exhibits at World's Fair," *Wall Street Journal*, March 31, 1939. For AT&T's own account, see Williams, "Our Exhibits."

62. Williams, "Our Exhibits," 80.

63. Williams, "Our Exhibits," 76–77.

64. Frank S. Adams, "Mrs. Roosevelt Heads Day's Throng of Visitors at the World's Fair," *New York Times*, May 16, 1939, 18; "Preview Is Given at Phone Exhibit: Guests of A. T. and T. Hear Voder," *New York Times*, April 28, 1939, 20; "Obliging Robots at Fair Staging Popularity Race," *Christian Science Monitor*, June 21, 1940, 2.

65. "R.U.R." *New York Times*, January 28, 1923, 136.

66. Karel Čapek, "We Alarm and Amuse M. Čapek," *New York Times Magazine*, May 16, 1926, 23.

67. Elektro was one in a lineage of robots built by Westinghouse engineers, beginning with Herbert Televox, built in 1927 by Roy Wensley; see Abnet, *American Robot*.

68. "Fair's New Robot Is Schizophrenic," *New York Times*, April 23, 1939.

69. Meyer Berger, "At the Fair," *New York Times*, May 5, 1939.

70. "Mechanical Man Rebels," *New York Times*, April 12, 1939.

71. Abnet, *American Robot*.

72. The first of these robots, Rastus, was made to look like a Black man, a fully racist embodiment of the idea of robot as slave.

73. Abnet, *American Robot*, 157.

74. R. L. Duffus, "The Beginning of a World, Not the End," *New York Times*, July 2, 1939, 88.

75. A letter from Bell to his parents describes the emperor only as saying, "I have heard—I have heard." Alexander Graham Bell, "Letter from Alexander Graham Bell to Alexander Melville Bell and Eliza Symonds Bell," June 27, 1876, manuscript/mixed material, https://www.loc

.gov/item/magbell.00500228/. In a 1930 promotional film, AT&T dramatizes the event: *The First Call*, AT&T Tech Channel video 13:11, https://techchannel.att.com/playvideo/2011 /08/31/AT&T-Archives-The-First-Call.

76. Douglas R. Hofstadter, *Fluid Concepts and Creative Analogies: Computer Models of the Fundamental Mechanisms of Thought* (New York: Basic Books, 1995), 157.

77. In 1992, the National Center for Voice & Speech, led by Dr. Ingo Titze, had created a software voice simulator called VoxInSilico to facilitate their research by enabling three-dimensional graphical modeling of the human vocal system by computer. Using the software, the team created a fully mathematical model of an operatic tenor's voice, which they named Pavaroboti and programmed to sing the high notes in a duet with Titze of "Nessun dorma" from *Turandot* by Giacomo Puccini, an aria that Luciano Pavarotti was famous for. The software was not available to the public and took specialized expertise in human vocal anatomy to use effectively. The Pavaroboti sample is far from perfect and took significant computing resources to generate, but it is still very impressive for the time. See demonstration videos at https://ncvs .org/why-voice-simulation/. The first voice synthesis software instruments to be released to the public were called Leon and Lola, from the sound effects company Zero-G in 2004. These synthesized voice libraries were based on the Vocaloid software, developed at the Pompeu Fabra University (Spain) as part of the dissertation project of Jordi Bonada with financing from the Yamaha Corporation. See Sarah A. Bell, "The dB in the .db: Vocaloid Software as Posthuman Instrument," *Popular Music and Society* 39, no. 2 (May 2016): 222–240. doi:10 .1080/03007766.2015.1049041.

CHAPTER 2

1. W. Koening, Hugh K. Dunn, and L. Y. Lacy, "The Sound Spectrograph," *Journal of the Acoustical Society of America* 18, no. 1 (1946): 19–49. doi:10.1121/1.1916342.

2. Homer Dudley, "Fundamentals of Speech Synthesis," *Journal of the Audio Engineering Society* 3, no. 4 (October 1955): 177. http://www.aes.org/e-lib/browse.cfm?elib=9.

3. On the prejudice of the Bells' oralism, see Katie Booth, *The Invention of Miracles: Language, Power, and Alexander Graham Bell's Quest to End Deafness* (New York: Simon & Schuster, 2021).

4. F. Barrows Colton, "Miracle Men of the Telephone," *National Geographic Magazine*, March 1947, 275.

5. Dudley, "Fundamentals," 184.

6. G. Oscar Russell, *The Vowel: Its Physiological Mechanism as Shown by X-Ray* (Columbus, OH: Ohio State University Press, 1928).

7. Hugh K. Dunn, "The Calculation of Vowel Resonances, and an Electrical Vocal Tract," *Journal of the Acoustical Society of America* 22, no. 6 (1950): 741.

8. Dunn, "The Calculation of Vowel Resonances," 742.

9. Dunn, "The Calculation of Vowel Resonances," 752.

10. Claude E. Shannon, "A Mathematical Theory of Cryptology," Bell Laboratories Memorandum MM-45-110–92, September 1, 1945.

11. Claude E. Shannon, "A Mathematical Theory of Communication," parts I and II, *Bell System Technical Journal* 27, no. 3 (July 1948): 379–423; Claude E. Shannon, "A Mathematical Theory of Communication," part III, *Bell System Technical Journal* 27, no. 4 (October 1948): 623–656. For a history of the evolution of the concepts in this paper, see Jimmy Soni and Rob Goodman, *A Mind At Play: How Claude Shannon Invented the Information Age* (New York: Simon & Schuster, 2017).

12. Bell also promoted their experimental mobile radiotelephones.

13. *Eye on Research*, "The Six Parameters of PAT" (London: BBC, 1958), uploaded March 26, 2012, Vimeo video, 30:23, https://vimeo.com/39186607.

14. Alison Taubman, "Talking Technology: How Machines Learned to Speak," National Museums Scotland, Edinburgh, https://blog.nms.ac.uk/2018/05/17/talking-technology-how-machines-learned-to-speak/.

15. D. W. Farnsworth, "High-Speed Motion Pictures of the Human Vocal Cords," *Bell Laboratories Record* 18 (March 1940): 203–208.

16. "Speech Machine Can Sing," *Science News Letter* 80 (November 25, 1961): 347.

17. Sarah A. Bell, "Federal Support for the Development of Speech Synthesis Technologies: A Case Study of the Kurzweil Reading Machine," *Information & Culture, A Journal of History* 58, no. 1 (January 2023): 39–65.

18. Will Lissner, "Computing 'Brain' Will 'Memorize,'" *New York Times*, December 12, 1947, 2, ProQuest Historical Newspapers.

19. Lissner, "Computing 'Brain' Will 'Memorize.'"

20. The computer was delivered in 1949 and began operating in 1951.

21. Lissner, "Computing 'Brain' Will 'Memorize.'"

22. Edmund C. Berkeley, *Giant Brains, or Machines That Think* (New York: John Wiley & Sons, 1949), x.

23. John B. Thurston, "Devaluing the Human Brain: *Cybernetics*, by Norbert Wiener," *Saturday Review*, April 23, 1949, 24.

24. Ronald R. Kline, *The Cybernetics Moment: Or Why We Call Our Age the Information Age* (Baltimore, MD: Johns Hopkins University Press, 2015), 69.

25. Wiener's warning from *Cybernetics* requires no mathematical competence to understand: "The first industrial revolution, the revolution of the 'dark satanic mills,' was the devaluation of the human arm by the competition of machinery. There is no rate of pay at which a United States pick-and-shovel laborer can live which is low enough to compete with the work of a steam shovel as an excavator. The modern industrial revolution is simply bound to devaluate the human brain at least in its simpler and more routine decisions. Of course, just as the

skilled carpenter, the skilled mechanic, the skilled dressmaker have survived in some degree the first industrial revolution, so the skilled scientist and the skilled administrator may survive the second. However, taking the second revolution as accomplished, the average human being of mediocre attainments or less has nothing to sell that it is worth anyone's money to buy." Norbert Wiener, *Cybernetics: Or Control and Communication in the Animal and the Machine* (Cambridge, MA: MIT Press, 2019 [1948]), 40.

26. John Kobler, "You're Not Very Smart After All," *Saturday Evening Post*, February 18, 1950, 25. Kobler's article also raises these stakes, as he discusses in "red scare" terms the Soviet development of computers, as well as US development in this area.

27. It should be noted that neither example is fictional; Mauchly did write a gin rummy program for a computer, and it was Wiener, along with some of the Macy conference cyberneticists, who theorized that both human brain "insanity" and electronic brain breakdown, were caused by rogue impulses in these systems.

28. This article, while fully utilizing the metaphor, also attempted to describe what computers could and could not do in 1950 with some accuracy. It is somewhat unique among journalistic accounts in that it includes an entire paragraph about the difficulties in maintaining the electronic components vulnerable to heat, dust, and other environmental givens. It also quotes IBM president Thomas J. Watson, Sr., as stating, "No machine can take the place of the scientist; this machine only leaves him more time for creative thinking," a now-familiar promise and the position that his son would double-down on when he became IBM president in the middle of the decade.

29. Paul E. Ceruzzi, *A History of Modern Computing*, 2nd ed. (Cambridge, MA: MIT Press, 2003).

30. US Bureau of the Census, "Section 31, 20th Century Statistics," *Statistical Abstract of the United States*, 119 (December 9, 1999): 885.

31. "TV Spot Biz Hits $116,935,000 for 1st Quarter in '57," *Variety*, May 29, 1957, 37.

32. Waldemar Kaempffert, "Science in Review," *New York Times*, June 8, 1952, E11. John von Neumann's team at Princeton's Institute for Advanced Study was also working on computerized weather forecasting with a machine nicknamed "Maniac" (after "ENIAC").

33. Computers might have been a difficult sell on camera but not behind it, as the networks salivated at the prospect of crunching viewership data to increase the value of television ad sales.

34. Technically, Franklin D. Roosevelt was the first US president to appear on television, from the New York World's Fair on April 30, 1939, but this appearance was broadcast only on receivers at the fairgrounds and at Radio City in Manhattan.

35. *Broadcasting*, "Network Reporting," November 10, 1952, 27.

36. *Broadcasting*, "Network Reporting," 27.

37. "CBS News Election Coverage: November 4, 1952," YouTube video, 31:02, https://www .youtube.com/watch?v=5vjD0d8D9Ec.

38. "CBS News Election Coverage."

39. "CBS Television Coverage of Election Returns Resulted in Landslide Victory for Network," *New York Times*, November 7, 1952, 31.

40. "CBS and NBC Electric TV Robots to Spot Election Night Trends," *Variety*, October 15, 1952, 1.

41. "Machine vs. Man," *Variety*, November 12, 1952, 25.

42. Joe Csida, "Web Coverage of Election Chaos Rates Bows for Entire Industry," *Billboard*, November 15, 1952, 3.

43. See John M. Jordan, *Machine-Age Ideology: Social Engineering and American Liberalism, 1911–1939* (Chapel Hill, NC: University of North Carolina Press, 1994) or, more recently, Sarah E. Igo, *The Averaged American: Surveys, Citizens, and the Making of a Mass Public* (Cambridge, MA: Harvard University Press, 2008).

44. "The Univac and the Unicorn," *Wall Street Journal*, October 17, 1952, 6.

45. Charles J. Swift, "Professor Univac," *Wall Street Journal*, November 14, 1952, 26.

46. Ahead of election night, some markets showed an episode of *Johns Hopkins Science Review*, during which UNIVAC cocreator John Mauchly addressed the question "Can Machines Think?" Dr. Mauchly was not inclined to personify the computer, instead stressing that the computer does "anything which the human being can direct that computer to do," only faster and without error. Available at https://www.youtube.com/watch?v=WdJWsexGupE.

47. Charles Collingwood, "My Life & Times with UNIVAC," *Variety*, January 5, 1955, 99.

48. John Morreall, "Philosophy of Humor," in Edward N. Zalta (ed.), *Stanford Encyclopedia of Philosophy* (fall 2020), https://plato.stanford.edu/archives/fall2020/entries/humor/.

49. Sam Chase, "Review Digest: Big 'Diamond Jubilee' Is Lacking in Carats," *Billboard*, November 6, 1954, 19; Lawrence Laurent, "They Had a Good One on the Co-ax this Time," *Washington Post and Times Herald*, October 26, 1954, 39.

50. Jack Gould, "Television in Review: 'Light's Jubilee' a Fine Salute to Edison," *New York Times*, October 25, 1954, 36.

51. George Rosen, "Selznick 'Lamp' Discovers America in $1,000,000 Fusion of Talents," *Variety*, October 27, 1954, 28.

52. "GE's Diamond Jubilee Special, Part 1," uploaded by Museum of Innovation and Science, March 15, 2012, YouTube video, 5:19, https://www.youtube.com/watch?v=A5nUSUNuO30.

53. "GE's Diamond Jubilee Special, Part 2," uploaded by Museum of Innovation and Science, March 15, 2012, YouTube video, 12:24, https://www.youtube.com/watch?v=PYJEzQpYJsY&t=625s.

54. "Linkletter Gag Cues Suit for $1,385,000," *Variety*, May 29, 1957, 37.

55. "Univac Takes a Bride," *Variety*, September 5, 1956, 23.

56. "Television Reviews: *People Are Funny*," *Variety*, September 19, 1956, 70.

57. *How to Marry a Millionaire*, season 1, episode 33, "The Truthivac," directed by Bernard Wiesen (Los Angeles: 20th-Century Fox, 1958).

58. The computer dating gone awry gag also shows up a decade later in an episode of the popular spy spoof *Get Smart*. In season 4, episode 18, "Absorb the Greek," the CONTROL agency chief is matched up by computer with a young Greek woman. It turns out that the computer matchmaking was all a ruse so that the Chief could meet covertly with the young woman, who is an informant. The CONTROL computer plays only a minimal role, but the fact that no one makes fun of or questions the legitimacy of computer matchmaking as a concept, only why the Chief would need a date, suggests a shift in public attitudes over the decade.

59. "Radio-Television: IBM Dickering 1st TV Entry," *Variety*, November 18, 1959, 28.

60. Nikil Saval, *Cubed: A Secret History of the Workplace* (New York: Doubleday, 2014).

61. Saval, *Cubed*, 158.

62. "FORTUNE 500, 1955 Full List," *CNN Money*, https://money.cnn.com/magazines/fortune/fortune500_archive/full/1955/1.html.

63. IBM Archives, "1955," *History of IBM*, https://www-03.ibm.com/ibm/history/history/year_1955.html.

64. Saval, *Cubed*, 157.

65. Harwood, *The Interface*, 163.

66. Harwood, *The Interface*. The idea that Watson directed Noyes to redesign everything from "curtains to computers" is a paraphrase from Harwood's Introductory chapter.

67. Remington Rand produced a film in 1960 that appears to roast, if not outright parody, *The Information Machine*. Rand's film, called *Information Age: Then and Now*, opens with the cartoon image of two cavemen whose heads are photos of the ENIAC creators J. Presper Eckert and John Mauchly. "In the beginning," these two cavemen "discovered they could count with their fingers" but quickly learn to stack rocks, draw hatch marks, and so forth, all "sound management decisions." Computer History Museum, YouTube video, 13:16, https://www.youtube.com/watch?v=h4wQJfdhOlU.

68. Charles Eames, *The Information Machine* draft of credits, 1957, box II: 149, folder 11, Projects: The Information Machine, Charles and Ray Eames Collection, Library of Congress.

69. IBM Archives, "Popularizing Math and Science," *IBM at 100*, https://www.ibm.com/ibm/history/ibm100/us/en/icons/mathandscience/.

CHAPTER 3

1. Cover art by Boris Artzybasheff, April 2, 1965.

2. "The Cybernated Generation," *TIME Magazine*, April 2, 1965, 85.

3. "Cybernated Generation," 85.

4. "Cybernated Generation," 88.

5. "Cybernated Generation," 90.

6. Ceruzzi, *History*, 110.

7. Caleb Pirtle, *Engineering the World: Stories from the First 75 Years of Texas Instruments* (Dallas, TX: Southern Methodist University Press, 2005).

8. Frederik Nebeker, *Signal Processing: The Emergence of a Discipline, 1948 to 1998* (Rutgers, NJ: State University of New Jersey, 1998), 10.

9. "Quantized" here refers to measuring any significant discrete change in quantity.

10. Jon W. Bayless, S. J. Campanella, and Aaron J. Goldberg, "Voice Signals: Bit-by-Bit," *IEEE Spectrum* 10, no. 10 (1973): 28; Bernard M. Oliver, John R. Pierce, and Claude E. Shannon, "The Philosophy of PCM," *Proceedings of the IRE* 36, no. 11 (1948): 1324–1331.

11. Quoted in Nebeker, *Signal Processing*, 6; full transcript of Nebeker conducted oral interview with Flanagan online at http://ethw.org/Oral-History:James_L._Flanagan.

12. Ronald E. Crochiere and James L. Flanagan, "Current Perspectives in Digital Speech," *Communications Magazine* 21, no. 1 (1983): 33. See also Sterne, *MP3*.

13. Oral history, quoted in Nebeker, *Signal Processing*, 72; see also James Kaiser, an oral history conducted in 1997 by Andrew Goldstein and Janet Abbate, IEEE History Center, Piscataway, NJ, USA, http://ethw.org/Oral-History:James_Kaiser.

14. James Kaiser, an oral history conducted in 1997 by Andrew Goldstein and Janet Abbate, IEEE History Center, Piscataway, NJ, http://ethw.org/Oral-History:James_Kaiser.

15. John L. Kelly and Carol C. Lochbaum, "Speech Synthesis," *Proceedings of the Fourth International Congress on Acoustics*, Copenhagen, 1962, Paper G42, 1–4.

16. Kelly and Lochbaum, "Speech Synthesis," 4.

17. Quoted in Jon Gertner, *The Idea Factory: Bell Labs and the Great Age of American Innovation* (New York: Penguin, 2013), 326.

18. Gertner, *Idea Factory*.

19. David G. Stork, ed., *HAL's Legacy: 2001's Computer as Dream and Reality* (Cambridge, MA: MIT Press, 1997).

20. Experiments in Art and Technology (EAT), "A Brief History and Summary of Major Projects, 1966–1998," March 1, 1998, https://www.vasulka.org/archive/Writings/EAT.pdf. See also https://www.bell-labs.com/about/history/innovation-stories/genesis-eat/#gref.

21. "Max Mathews, 84, Pioneer in Making Computer Music," *New York Times*, April 24, 2011, 22.

22. Timothy D. Taylor, "The Avant-Garde in the Family Room: American Advertising and the Domestication of Electronic Music in the 1960s and 1970s," in Trevor Pinch and Karin

Bijsterveld (eds.), *The Oxford Handbook of Sound Studies* (New York: Oxford University Press, 2012), 389.

23. Carlos estimated that the project took over 1,100 hours to complete. Amanda Sewell, *Wendy Carlos: A Biography* (New York: Oxford University Press, 2022).

24. The RIAA's certification database can be searched at https://www.riaa.com/gold-platinum/. Wendy Carlos's Grammy Award information can be found at at https://www.grammy.com /artists/wendy-carlos/1220.

25. M. Thieberger and Charles Dodge, "An Interview with Charles Dodge," *Computer Music Journal* 19, no. 1 (spring 1995): 11.

26. Thieberger and Dodge, "An Interview with Charles Dodge," 14–15.

27. Thieberger and Dodge, "An Interview with Charles Dodge," 15. The well-known computer music composer and software developer Miller Puckette said of *Speech Songs*:

    By the 1970s, the listening public was well accustomed to hearing human voices emerge from loudspeakers, and a listener could identify the sound of a reproduced, prerecorded voice as "real," but replacing this voice with a clearly artificial one put the question of liveness, and of naturalness, back into conscious play in an act of "making it strange." This impression was enhanced by the crudeness of the sounds available at the time. The most important measure of speech quality to researchers at Bell Labs was intelligibility (as opposed to audio quality or classically defined beauty), and the hyperintelligible but metallic machine voice of *Speech Songs* places the listener in an unfamiliar relationship with the music. In exploiting the poetics of this situation, Dodge presages, for instance, Laurie Anderson's *O Superman* and, arguably, the widespread use of synthesized and "autotuned" singing voices in popular music at present.

    Miller Puckette, "The Contributions of Charles Dodge's *Speech Songs* to Computer Music Practice," in Miller Puckette and Karry L. Hagan (eds.), *Between the Tracks: Musicians on Selected Electronic Music* (Cambridge, MA: MIT Press, 2020), chapter 4, https://doi.org/10 .7551/mitpress/12039.003.0005. Drawing out the influence of voice synthesis development on experimental and popular music is outside the scope of this book, but it is an important point for cultural analysis of voice synthesis technologies, and for which there is a developing body of interesting scholarship.

28. Bishnu S. Atal, "The History of Linear Prediction," *IEEE Signal Processing Magazine*, March 2006, 154–161.

29. Bishnu S. Atal and Suzanne Hanauer, "Speech Analysis and Synthesis by Linear Prediction of the Speech Wave," *Journal of the Acoustical Society of America* 50, no. 2B (1971): 637–655.

30. Peter Elias, "Predictive Coding I," *IRE Transactions in Information Theory*, IT-1, no. 1 (March 1955): 6–24; Peter Elias, "Predictive Coding II," *IRE Transactions in Information Theory*, IT-1, no. 1 (March 1955): 24–33.

31. Homer W. Dudley, "The Carrier Nature of Speech," *Bell System Technical Journal* 19, no. 4 (1940): 495–515.

32. Nebeker, *Signal Processing*, 63.

33. Atal, "History of Linear Prediction," 155.

34. Nebeker, *Signal Processing*, 65.

35. I maintain the gendered pronoun in use at the time for clarity and consistency with the technical literature.

36. This is the argument of the sociologist Daniel Bell, who is often cited as providing an early definition of the Information Age by arguing that "a post-industrial society is, preeminently, a high technology economy," in "The Coming of the Post-Industrial Society."

37. Stacy V. Jones, "Patent Awarded to Palsy Victim," *New York Times*, Saturday, May 16, 1964, 28. See also Charlotte R. Burmeister, "President's Committee on the Employment of the Handicapped," *Workplace Health and Safety* 10, no. 9 (1962): 29. The honor was given to the person "who in the past calendar year has surmounted his or her handicap to become a useful citizen and who has helped encourage, inspire, or facilitate the employment of other handicapped individuals," a statement that conflates citizenship with employment, or economic usefulness.

38. "Man's Breath Runs Typewriter," *Life*, December 1, 1952, 77–80.

39. Eric Northrup, "Electronics Wizard," *Mechanix Illustrated*, June 1953, 61. A profile in *Life* magazine in 1962 stated that Avakian hoped someday to use computers to discover "faulty circuits" in the brain of palsy victims that could then be removed with radiation. See "Proud Win for a Man with a Will," *Life*, May 11, 1962, 56.

40. Vartanig G. Vartan, "Talking Computer Quotes Stock," *New York Times*, Monday, May 11, 1964, 47.

41. Vartan, "Talking Computer Quotes Stock," 49.

42. Devin Kennedy, "The Machine in the Market: Computers and the Infrastructure of Price at the New York Stock Exchange, 1965–1975," *Social Studies of Science* 47, no. 6 (2017): 890.

43. Unionized employees were a specific target of automation in this industry, as in many others.

44. Kennedy, "New York Stock Exchange," 894 (emphasis added).

45. Kennedy, "New York Stock Exchange," 891; the richness of Kennedy's evidence can only be glossed over here, but it shows how NYSE managers *chose* not to deploy technology to monitor brokers even though the Securities and Exchange Commission (SEC) pressed them to do so, even going so far as to claim that this was not technically feasible, as it would result in *costs* in speed and accuracy. Kennedy points out that the SEC's willingness to trust this evaluation by the NYSE reflects aspects of the broader politics of markets, technology, and regulation in the US during this period.

46. Vartanig G. Vartan, "Automation Gain Noted in Market," *New York Times*, May 17, 1964, 183, 197.

47. Kennedy, "New York Stock Exchange," 894.

48. G. E. DuBois, "Computer Voice Output Device," Technical Report 00.937, IBM Data Systems Division Development Laboratory, Poughkeepsie, NY, December 15, 1962, collection 417, box 6, Speech Synthesis History Project, Smithsonian National Museum of American History archives (hereafter cited as SSHP).

49. In actuality, they were timed to 294 ms, leaving the 6 ms required for control purposes.

50. DuBois, Technical Report, 2.

51. "Compatible," *New Yorker*, June 12, 1965, 29–30.

52. A. Bruce Urquhart, "Voice Output from IBM System/360," Technical Report 21.143, IBM Data Systems Division, Kingston, NY, February 5, 1965, collection 417, box 6, SSHP. Peripherals were the 7770 and 7772 Audio Response Units. The 7770 stored the vocabulary on a drum as described; the 7772 stored the vocabulary digitally on a disk file.

53. Gene Smith, "Chatting via Computer," *New York Times*, September 12, 1971.

54. Gene Gleason, "I.B.M. Device 'Knows' 10 Languages," *New York Herald Tribune*, February 14, 1958, A8; some of the newspaper coverage perpetuated some misinformation about the computer's translation abilities, as Ceruzzi notes in *History of Modern Computing*.

55. John McCarthy, Marvin L. Minsky, Nathaniel Rochester, and Claude E. Shannon, "A Proposal for the Dartmouth Summer Research Project on Artificial Intelligence," August 31, 1955. Reprinted in *AI Magazine* 27, no. 4 (2006): 12–14.

56. Alan M. Turing, "Computing Machinery and Intelligence," *Mind* 59, no. 236 (October 1950): 433–460.

57. James Hawthorne, "A Bad Case of Buttonitus: *The Jetsons*, Animation and the Enlightenment Idea of Progress," *Static* 1, no. 9 (2012): 1–8. http://static.londonconsortium.com/issue09.

58. Brian Cowlishaw, "No Future Shock Here: The Jetsons, Happy Tech, and the Patriarchy," in R. C. Neighbors and Sandy Rankin (eds.), *The Galaxy Is Rated G: Essays on Children's Science Fiction Film and Television* (Jefferson, NC: McFarland & Co., 2011), 183–193.

59. *The Computer Wore Tennis Shoes*, directed by Robert Butler (Los Angeles: Walt Disney Productions, 1969).

60. Disney made only three animated features during the 1960s, but it more than doubled the number of live-action features from the previous decade, releasing forty-seven (not including the decade's highlight, the Oscar-winning 1964 movie *Mary Poppins*, which was a live-action film with animated and hybrid animation and live-action sequences).

61. "The Computer Wore Tennis Shoes," *Variety*, December 31, 1969.

62. Edwards, *Closed World*, 7. See the text for Edwards's full analysis of closed world/green world examples in 1960s–1980s era science fiction.

63. Edwards, *Closed World*, 314.

64. Kubrick's black comedy *Dr. Strangelove* was his other Cold War movie, of course.

65. In the significant body of analyses of the film, many critics discuss the symbolic nature of the environment of the Discovery One, and HAL within it, including Adam Roberts, who calls it a fully realized environment that rejects human existence. He credits Kubrick with forging a Space Age visual lexicon for film that brought new seriousness and aesthetic respectability to visual science fiction. Adam Roberts, *The History of Science Fiction* (New York: Palgrave Macmillan, 2016). Liz Faber does an extensive psychoanalytic reading of the ship as both womb and phallus in their study of computer voices in science fiction. Liz Faber, *The Computer's Voice: From Star Trek to Siri* (Minneapolis, MN: University of Minnesota Press, 2020).

66. Spock and McCoy, who are often at odds, represent an obvious bifurcation of mind and body, with Spock telling Kirk what he should be thinking and McCoy advising him about his feelings. Interestingly, Spock and McCoy also "share" a female assistant—the actress Majel Barrett, who played both Head Nurse Christine Chapel *and* the voice of the *Enterprise* computer. Barrett/Chapel isn't presented as the voice of the computer in the context of the series; indeed, when an episode featured the computer voice, Nurse Chapel didn't appear in it at all (presumably so that audiences wouldn't make the connection), but the psychology of series creator Gene Roddenberry's own real-life girlfriend, Barrett, being "split," body and mind, between Captain Kirk's heart and brain advisors is nevertheless compelling.

67. *Star Trek*, season 1, episode 19, "Tomorrow Is Yesterday," directed by Michael O'Herlihy (Culver City, CA: Desilu Studios, 1967).

CHAPTER 4

1. Peter J. Schuyten, "Electronic Games a Big Winner for the Holidays," *New York Times*, November 15, 1979, D1.

2. For histories of educational computing in the 1970s, see Joy Lisi Rankin, *A People's History of Computing* (Cambridge, MA: Harvard University Press, 2018) and in the UK, Neil Selwyn, "Learning to Love the Micro: The Discursive Construction of 'Educational' Computing in the UK, 1979–89," *British Journal of Sociology of Education* 23, no 3 (2002): 427–443; for a history of precomputer teaching machines, see Audrey Watters, *Teaching Machines: The History of Personalized Learning* (Cambridge, MA: MIT Press, 2021); for ideas about educational technology contemporary to the 1970s, see Seymour Papert, *Mindstorms: Children, Computers and Powerful Ideas* (New York: Basic Books, 1980); and for analysis, see Morgan G. Ames, *The Charisma Machine: The Life, Death, and Legacy of One Laptop per Child* (Cambridge, MA: MIT Press, 2019).

3. TI press release, posted at Jeorg Woerner's *Datamath Calculator Museum*, http://www.datamath.org/Story/Intel.htm#Press%20release%20TI%201974. See also http://www.datamath.org/Story/Intel.htm#TMS1000.

4. Paul Breedlove, laboratory and engineering notebook, March 1, 1976, RG-6 Semiconductor Group, 88-67 Speak & Spell, Texas Instruments records, DeGolyer Library, Southern Methodist University, University Park, TX.

5. This group was directed by George Doddington, who had completed his PhD thesis in 1970 by conducting experiments in speaker recognition with James Flanagan at Bell Labs. Doddington's small team was working on computer interfaces for an experimental speaker recognition unit under a contract with the US military.

6. "How the IDEA System Launched One of TI's Most Popular Consumer Products," *AusTInews*, February 1980, 9, with Breedlove's notes attached, RG-6 Semiconductor Group, 88-67 Speak & Spell Texas Instruments records, DeGolyer Library, Southern Methodist University, University Park, TX.

7. Gene Frantz, *The Speak N Spell* (Houston, TX: Connexions, 2013), 23.

8. Richard A. Shaffer, "Electronic Gadgets That Can Talk, Spell May Open New Path for Consumer Goods," *Wall Street Journal*, June 12, 1978, 12.

9. "New TI Learning Aid 'Talks' to TV Audience," *DallaSite*, August 1978, RG-6 Semiconductor Group, 88-67 Speak & Spell Texas Instruments records, DeGolyer Library, Southern Methodist University, University Park, TX.

10. Both companies have ties to Bell though. Shockley, one of the inventors of the transistor at Bell left to form his own company. In a move that some describe as a mutiny, eight of Shockley's best, including Noyce, left his company to form Fairchild. Noyce eventually founded Intel with Gordon Moore. Also, Gordon Teal, Kilby's boss at TI, had started out in the Shockley lab at Bell and is seen as one of the key people responsible for the success of TI in the consumer electronics market.

11. Gene Frantz and Richard Wiggins, "The Development of 'Solid State Speech' Technology at Texas Instruments," *Acoustics, Speech, and Signal Processing Newsletter* 53, no. 1 (1981): 34.

12. Votrax advertisement, *Scientific American*, September 1972, 165.

13. Eric J. Lerner, "Products That Talk: Speech-Synthesis Devices Are Being Incorporated into Dozens of Products as Difficult Technical Problems Are Solved," *IEEE Spectrum* 19, no. 7 (1982): 32.

14. Richard Wiggins and Larry Brantingham, "Three-Chip System Synthesizes Human Speech," *Electronics*, August 31, 1978, 109–116.

15. Robert Gannon, "Success in the Quest for Machines That Talk . . . and Listen," *Popular Science*, August 1980, 56.

16. Wiggins and Brantingham, "Three-Chip System," 110.

17. Richard H. Wiggins and George L. Brantingham, Speech Synthesis Integrated Circuit Device. US Patent 4,209,836, filed April 28, 1978, and issued June 24, 1980, 154–156.

18. Frantz, *Speak N Spell*, 24.

19. Frantz, *Speak N Spell*, 25.

20. Frantz, *Speak N Spell*, 49.

21. Frantz, *Speak N Spell*, 29–30.

22. Frantz, *Speak N Spell*, 28.

23. Frantz, *Speak N Spell*, 30.

24. Frantz and Wiggins, "Solid State Speech," 36.

25. Frantz, *Speak N Spell*, 40.

26. Frantz, *Speak N Spell*, 38.

27. Frantz, *Speak N Spell*, 35.

28. Frantz, *Speak N Spell*, 36.

29. Frantz, *Speak N Spell*, 46.

30. Paul S. Breedlove and James H. Moore, Electronic Learning Aid or Game Having Synthesized Speech. US Patent 4,516,260, filed August 29, 1980, and issued May 7, 1985.

31. Noam Chomsky, *Syntactic Structures* (Berlin: Walter de Gruyter, 1957). Although Chomsky doesn't explicitly label linguistic structures as innate in this work, his separation of syntax and semantics had a significant impact on the development of computer science. For criticisms of universal grammar, see Michael Tomasello, *Origins of Human Communication* (Cambridge, MA: MIT Press, 2008).

32. Comments from focus groups, summer 1977, RG-6 Semiconductor Group, 88-61 Speak & Spell, Texas Instruments records, DeGolyer Library, Southern Methodist University, University Park, TX.

33. Frantz, *Speak N Spell*, 42.

34. A presentation that Breedlove gave for the Consumer Calculator Division three months before the focus groups contains artist's renderings of five character options: a bee, a typewriter, a parrot called Wordy Birdy, the computer, and a male superhero. The computer is depicted with round eyes resembling tape reels and the display was its "mouth." Paul Breedlove, presentation package for Speak & Spell program, February 28, 1977, RG-6 Semiconductor Group, 88-61 Speak & Spell Texas Instruments records, DeGolyer Library, Southern Methodist University, University Park, TX.

35. Frantz, *Speak N Spell*, 42.

36. Frantz and Wiggins, "Solid State Speech," 36.

37. C. D. Terrell and O. Linyard, "Evaluation of Electronic Learning Aids: Texas Instruments' 'Speak & Spell,'" *International Journal of Man-Machine Studies* 17, no. 1 (1982): 65.

38. For a comprehensive and rich discussion of postwar play, see Amy F. Ogata, *Designing the Creative Child: Playthings and Places in Midcentury America* (Minneapolis, MN: University of Minnesota Press, 2013).

39. A *TIME* magazine poll in December 1982 found that 68 percent of adults believed that home computers would improve the quality of their children's education. As there were fewer than 3 million home computers in circulation in 1982, this shows that people were anticipating

what the machines would be useful for. Otto Friedrich, "The Computer Moves In," *TIME*, January 3, 1983.

40. Carroll Pursell, *From Playgrounds to PlayStation: The Interaction of Technology and Play* (Baltimore, MD: Johns Hopkins University Press, 2015), 1.

41. Pursell, *Playgrounds*, 9.

42. Brian Santo, "25 Microchips That Shook the World," *IEEE Spectrum*, May 1, 2009, https://spectrum.ieee.org/tech-history/silicon-revolution/25-microchips-that-shook-the-world.

43. STEM, the educational acronym for "science, technology, engineering, and mathematics," was introduced by the National Science Foundation (NSF) in 2001. Judith Ramaley, assistant director of education at NSF, rearranged the previous acronym, SMET.

44. "Electronic Toys Becoming Popular Games Creating Multi-Million-Dollar Market," *Playthings*, January 1977, 33.

45. Advertisement, *Playthings*, February 1977, 180.

46. Advertisement, *Playthings*, 152.

47. Advertisement, *Playthings*, 182.

48. Advertisement, *Playthings*, 335.

49. Advertisement, *Playthings*, 225.

50. Advertisement, *Playthings*, 237.

51. Lynn Langway, "Turned-on Toys," *Newsweek*, December 11, 1978, 78.

52. Advertisement, *Playthings*, 349.

53. Advertisement, *Playthings*, 357.

54. "What's Selling," *Playthings*, January 1978, 16–17. This happened in spite of the fact that the small company Kenner had only gotten a handshake agreement to create a *Star Wars* line a few months before. Action figures, still working through the production process overseas, were presold at Christmas with the promise of a spring 1978 delivery. *The Toys That Made US*, season 1, episode 1, "Star Wars," directed by Tom Stern (Los Angeles: Nacelle Company, 2017).

55. Langway, "Turned-on Toys," 76.

56. Langway, "Turned-on Toys," 76.

57. Victor K. McElheny, "New Markets Are Sought for Miniaturized Computers," *New York Times*, January 16, 1975, 74.

58. Langway, "Turned-on Toys," 79.

59. Michael J. Freedman, Real Time Conversational Toy. US Patent 4,078,316, filed June 24, 1976, and issued March 14, 1978; a later version from Tiger Electronics in the 1990s included voice synthesis, but the 1978 original did not.

60. Advertisement, *Playthings*, June 1978.

61. "Parker Brothers Receiving Questionnaire Responses," *Playthings*, January 1979, 98.

62. Langway, "Turned-on Toys," 78; although Speak & Spell gets a brief mention in the *Newsweek* article, it was not one of the products that the kid panel "tested." This might have been due to production schedules, but it isn't addressed in the article. Simon, the much less expensive family game, which was famously promoted at an exclusive party held at the disco club Studio 54 in New York City in spring 1978, was to be the hottest toy that Christmas. Simon, designed by Ralph Baer, used TI's TMS1000 microprocessor.

63. Tom Murn, "Model Kits—the Staple of Life," *Playthings*, April 1979, 25.

64. "Electronics: Big Entry into Market by Outside Firms," *Playthings*, April 1979, 26.

65. Diane McWhorter, "Electronic Shock in Toyland," *Boston Magazine*, October 1978, 158.

66. Langway, "Turned-on Toys," 77.

67. The rollercoaster fortunes of the arcade and home gaming companies, especially Atari, is well documented elsewhere. For scholarly treatments of the social impacts, see Michael Z. Newman, *Atari Age: The Emergence of Video Games in America* (Cambridge, MA: MIT Press, 2017); Mark J. P. Wolf, ed., *Before the Crash: Early Video Game History* (Detroit, MI: Wayne State University Press, 2012); and for a cultural history, see Carly A. Kocurek, *Coin-Operated Americans: Rebooting Boyhood at the Video Game Arcade* (Minneapolis, MN: University of Minnesota Press, 2015).

68. Watters, *Teaching Machines*.

69. Watters, *Teaching Machines*, 44.

70. Watters, *Teaching Machines*, 45.

71. Watters, *Teaching Machines*, 80.

72. Watters, *Teaching Machines*, 97.

73. Watters, *Teaching Machines*, 257.

74. "Chicago U. Psychologist Tours for Tiger Products," *Playthings*, January 1981, 30.

75. Howard Gardner, "Toys with a Mind of Their Own," *Psychology Today*, November 1979, 94 (emphasis added).

76. Gardner, "Mind of Their Own," 98.

77. Paula Smith, "The Impact of Computerization on Children's Toys and Games," *Journal of Children in Contemporary Society* 14, no. 1 (1981): 73. Note that although published in an academic journal, the essay is an opinion piece.

78. Smith, "Impact," 81.

79. Smith, "Impact," 79.

80. McWhorter, "Electronic Shock," 170.

81. Isaac Asimov, "Video Games Are Dead: Long Live the Supergames of Tomorrow," *Video Review*, May 1983, 34.

82. Sherry Turkle, *The Second Self: Computers and the Human Spirit* (Cambridge, MA: MIT Press, 2005 [1984]), 24.

83. Turkle, *Second Self*, 22.

84. Armagnac, "Pedro," 72.

85. Turkle, *Second Self*, 18.

86. Turkle, *Second Self*, 19.

87. One of the best studies that shows this other strand of the metaphor is Edwards, *Closed World*. David Golumbia also discussed the impact of the language of computationalism at length in *The Cultural Logic of Computation* (Cambridge, MA: Harvard University Press, 2009).

88. *Computers Are People, Too!* directed by Denis Sanders (Los Angeles: Walt Disney Productions, 1982).

89. *TIME* changed the name of the citation to the gender-neutral "Person of the Year" in 1999. They had previously made awards for "Woman of the Year" and groups that included "Man and Woman of the Year."

90. On social reaction to *WarGames*, see Stephanie Ricker Schulte, "'The *WarGames* Scenario': Regulating Teenagers and Teenaged Technology (1980–1984)" *Television & New Media* 9, no. 6 (2008): 487–513.

CHAPTER 5

1. In 1960, psychiatrist Nathan S. Kline and polymath Manfred Clynes coined the term "cyborg," a portmanteau of "cybernetic" and "organism," or a self-regulating man-machine system, that they argued was necessary for traveling the long distances that manned space-flight would require. Manfred E. Clynes and Nathan S. Kline, "Cyborgs and Space," *Astronautics* 14, no. 9 (1960): 26–27.

2. Donna Haraway, "Manifesto for Cyborgs: Science, Technology, and Socialist Feminism in the 1980s," *Socialist Review* 80 (1985): 65–108.

3. Turkle, *Second Self*.

4. "This Day in History, January 25, 1979," Computer History Museum, https://www.computerhistory.org/tdih/january/25/.

5. Megan Prelinger, *Another Science Fiction: Advertising the Space Race 1957–1962* (New York: Blast Books, 2010), 197.

6. Klint Finley, "Tech Time Warp of the Week: Return to 1974 When a Computer Ordered a Pizza for the First Time," *Wired*, January 30, 2015, https://www.wired.com/2015/01/tech-time-warp-pizza/.

7. The subheadline for the 2015 *Wired* article doesn't even get Sherman's first name correct, referring to him as "John" rather than Donald. John was the name of one of the system's

designers, a further confusion of agency about who and what is "speaking" in the interaction, even if represented in a copy-editing mistake.

8. Mara Mills, "Hearing Aids and the History of Electronics Miniaturization," *IEEE Annals of the History of Computing* 33, no. 2 (2011): 24–45; "On Disability and Cybernetics: Helen Keller, Norbert Wiener, and the Hearing Glove," *Differences* 22, no. 2–3 (2011): 74–111; "Deafening: Noise and the Engineering of Communication in the Telephone System," *Grey Room* 43 (2011): 118–143.

9. Sarah S. Jain, "Enabling and Disabling the Prosthesis Trope," *Science, Technology, & Human Values* 24, no. 1 (1999): 31–54.

10. For a good summary of transhumanism, see Mark O'Connell's journalistic account *To Be a Machine* (New York: Doubleday, 2017).

11. Frances A. Koestler, "Chapter 21: Little Things That Make a Big Difference," in *The Unseen Minority: A Social History of Blindness in the United States* (Arlington, VA: American Foundation for the Blind, 2004), https://www.afb.org/online-library/unseen-minority-0/chapter-21.

12. F. S. Cooper, J. H. Gaitenby, and P. W. Nye, "Evolution of Reading Machines for the Blind: Haskins Laboratories' Research as a Case History," *Journal of Rehabilitation Research and Development* 21, no. 1 (1984): 51–87. See also "Optophones and Musical Print," January 5, 2015, *Sounding Out!* blog, https://soundstudiesblog.com/2015/01/05/optophones-and-musical-print/.

13. Cooper et al., "Evolution of Reading Machines for the Blind," 66.

14. Haskins researchers were able to build a computer-controlled formant synthesizer in 1966. Previous to this, they had been using custom analog circuitry etched by hand in the lab.

15. Cooper et al., "Evolution of Reading Machines for the Blind," 66.

16. "Overview of RLE Speech Research," *RLE Currents* 1, no. 1, December 1987, https://www.rle.mit.edu/media/currents/1-1.pdf.

17. Memo to David Maxey, October 25, 1990, collection 417, box 9, MIT folder, SSHP.

18. Jonathan Allen, M. Sharon Hunnicutt, and Dennis Klatt, *From Text to Speech: The MITalk System* (Cambridge, UK: Cambridge University Press, 1987).

19. Dennis Klatt, "The Klatt-Talk KT-1 Phonemic Synthesizer," preliminary draft, April 1, 1980, collection 417, box 8, MIT folder, SSHP.

20. Klatt, "The Klatt-Talk KT-1 Phonemic Synthesizer," draft, 2.

21. Klatt, "The Klatt-Talk KT-1 Phonemic Synthesizer," draft, 4.

22. Forest S. Mozer, Method and Apparatus for Speech Synthesizing. US Patent 4,214,125, filed January 21, 1977, issued July 22, 1980; Mozer also licensed the design to National Semiconductor, which used it in its DigiTalker synthesizer.

23. Letter from Gabriel F. Groner to H. David Maxey, January 30, 1989, collection 417, box 9, Telesensory Systems folder, SSHP. The KRM-IV European language versions introduced in 1985 used the Infovox SA-101 synthesizer from Infovox S.A. of Sweden and supported optical character recognition and text-to-speech in English, French, German, Italian, Spanish, and Swedish. The Kurzweil Personal Reader, introduced in April 1988, used DECtalk speech synthesis that also had been developed from the MIT system.

24. Letter from Paul Liniak to H. David Maxey, January 22, 1979, collection 417, box 9, Telesensory Systems folder, SSHP.

25. Letter from John E. Dennis to Michael Hyland, February 23, 1983, collection 417, box 9, Telesensory Systems folder, SSHP.

26. Telesensory Speech Systems brochure, 1982, collection 417, box 9, Telesensory Systems folder, SSHP.

27. Telesensory Speech Systems brochure, 1982, collection 417, box 9, Telesensory Systems folder, SSHP.

28. Allen et al., *From Text to Speech*, 7.

29. Memo from Frank Spitznogle to George Heilmeier, July 17, 1981, RG-6 Semiconductor Group, 88-61 Speak & Spell, Texas Instruments records, DeGolyer Library, Southern Methodist University, University Park, TX.

30. "ESA Honoured to Take Part in Hawking Tribute," June 15, 2018, https://www.esa.int/About_Us/Art_Culture_in_Space/ESA_honoured_to_take_part_in_Hawking_tribute.

31. Paul Sandle, "Stephen Hawking's Voice Was His Tool and His Trademark," Science News Reuters, March 14, 2018, https://www.reuters.com/article/idUSKCN1GQ2XC/.

32. Martin Rees, "Obituary, Stephen Hawking (1942–2018)," *Nature* 555 (March 22, 2018): 444, https://www.nature.com/articles/d41586-018-02839-9.

33. The UK first edition also features a photograph of Hawking in the same layout, but the photo is of Hawking, smiling, in front of a chalkboard full of equations. The tenth-anniversary edition features a stylized image of Hawking in front of an artist's rendering of a night sky, but no wheelchair.

34. Joao Medeiros, "How Intel Gave Stephen Hawking a Voice," *Wired*, January 13, 2015, https://www.wired.com/2015/01/intel-gave-stephen-hawking-voice/. The combination of components that made up the system that allowed Hawking to communicate with others was often collapsed in the press as "Hawking's voice" or "the synthesizer," even though the actual synthesizer was only one of several technologies that had to be made to work together. One academic paper incorrectly identifies the synthesizer by the name of another piece of software, Equalizer.

35. Letter from H. David Maxey to Dennis Klatt, April 16, 1988, collection 417, box 32, folder 6, SSHP.

36. Note from Dennis Klatt to H. David Maxey, April 29, 1988, collection 417, box 32, folder 6, SSHP.

37. Jason Fagone, "The Silicon Valley Quest to Preserve Stephen Hawking's Voice," *San Francisco Chronicle*, March 18, 2018, https://www.sfchronicle.com/bayarea/article/The-Silicon-Valley-quest-to-preserve-Stephen-12759775.php.

38. Ryan Parker, "Stephen Hawking Had a Single Joke Request When He Appeared on 'The Simpsons,'" *Hollywood Reporter*, March 14, 2018, https://www.hollywoodreporter.com/tv/tv-news/stephen-hawking-had-a-single-joke-request-he-appeared-simpsons-1094383/.

39. Ben Child, "Stephen Hawking: I'd Love to Play a Bond Baddie," *The Guardian*, December 2, 2014, https://www.theguardian.com/film/2014/dec/02/stephen-hawking-bond-baddie-film-theory-of-everything.

40. Mason was the husband of Hawking's nurse, Elaine Mason. Hawking and Elaine fell in love, eventually leaving their respective spouses to marry each other in 1995 (although Hawking had been separated from his first wife since 1990 and the couple were purportedly approving of one another's affairs). John Durham Peters observes that this relationship "closes [Hawking's] machine voice in a typically strange amorous circle," but one with unhappy consequences, as Hawking and Elaine divorced after eleven years, with Hawking pursuing the affections of another caregiver. Hawking's adult children accused Elaine of abusing Hawking, which he denied. Police investigated the accusations, but no criminal charges were ever filed.

41. Stephen Hawking, "The Computer," Internet Archive Wayback Machine, captured January 10, 2012, http://www.hawking.org.uk/the-computer.html.

42. Fagone, "Silicon Valley Quest."

43. Aly Weisman, "Stephen Hawking Gave Filmmakers a Priceless Gift After Watching the New Movie About His Life," *Business Insider*, October 30, 2014, https://www.businessinsider.com/stephen-hawking-voice-in-the-theory-of-everything-2014-10.

44. Peter Benie, "This Cambridge Life," April 26, 2018, https://medium.com/this-cambridge-life/the-man-who-helped-to-preserve-stephen-hawkings-iconic-voice-ee9e67e4c0e1.

45. Andrew Pollack, "Audiotex: Data by Telephone," *New York Times*, January 5, 1984, D2.

46. Rachel Handley, "Stephen Hawking's Voice, Made by a Man Who Lost His Own," *Beyond Words*, July 15, 2021, https://beyondwords.io/blog/stephen-hawkings-voice.

47. British Broadcasting Company, "Klatt's Last Tapes," audio program, 28:00, August 9, 2019, https://www.bbc.co.uk/programmes/b03775fy.

48. Paul Kemmerling, Richard Geiselhart, David E. Thorburn, and James Gary Cronburg, "A Comparison of Voice and Tone Warning Systems as a Function of Task Loading," ASD-TR-69-104, National Technical Information Service, Springfield, VA, September 1969.

49. David E. Thorburn, *Voice Warning Systems—A Cockpit Improvement That Should Not Be Overlooked* (Springfield, VA: National Technical Information Service, 1971), AMRL-TR-70-138.

50. Steve Tupper, "Kim Crow, the Original Bitching Betty," *Airspeed*, February 16, 2014, http://airspeedonline.com/2014/02/kim-crow-the-original-bitching-betty-audio-episode-show-notes/. This rationale for using a female voice may be folklore, as many air traffic and

military radio controllers in the 1970s were women, and women flew other kinds of military aircraft, as they had done since World War I. Later fighter jets built by Boeing used recordings made by an employee, Leslie Shook, but the "Bitching Betty" name continued. Royal Air Force pilots refer to the Voice Warning System as "Nagging Nora." Pilots have reported feeling quite attached to the voice in the specific aircraft that they flew.

51. Tupper, "Kim Crow."

52. "Betty, n." *Green's Dictionary of Slang*, online edition, https://greensdictofslang.com/entry /dx6wfha.

53. Paul Woodford, "F-15 Pilot Shares the History of 'Bitchin' Betty,'" *Hush-Kit, the Alternative Aviation Magazine*, February 5, 2018, https://hushkit.net/2018/02/05/f-15-pilot-shares-the -history-of-bitchin-betty/. The name of the woman who did the B-58 recordings was Joan Elms.

54. Jeff Schogol, "Woman Behind the F/A-18's 'Bitchin Betty' Voice Retires," *MarineCorps Times*, March 14, 2016, https://www.marinecorpstimes.com/news/your-marine-corps/2016/03/14 /woman-behind-the-f-a-18-s-bitchin-betty-voice-retires/.

55. Schogol, "'Bitchin Betty' Voice Retires."

56. Dennis Klatt, "Review of Text-to-Speech Conversion for English," *Journal of the Acoustical Society of America* 82 (1987): 784.

57. Dennis Klatt, "Speech Processing Strategies Based on Auditory Models," in R. Carlson and B. Granström (eds.), *The Representation of Speech in the Peripheral Auditory System* (New York: Elsevier, 1982), 183.

58. Inger Karlsson, "Female Voices in Speech Synthesis," *Journal of Phonetics* 19 (1991): 114.

59. One count of the phonetic literature found that 40.5 percent of studies had assembled only male speakers, with the vast majority of the rest using more male than female or an even number of male and female participants. Less than 5 percent of studies focused on female speakers. Caroline Henton, "Where Is Female Synthetic Speech?" *Journal of the International Phonetic Association* 29, no. 1 (1999): 51–61.

60. Nina Power, "Soft Coercion, The City, and the Recorded Female Voice," in Matthew Gandy and B. J. Nilsen (eds.), *The Acoustic City* (Berlin: JOVIS Verlag, 2014), 23–26.

61. Steven Leveen, "'Technosexism,'" *New York Times*, November 12, 1983, 23.

62. Leveen, "Technosexism."

63. Leveen's is the earliest articulation of this view in print that I have come across.

64. James R. Beniger and Clifford I. Nass, "Preprocessing: Neglected Component of Sociocy-bernetics," in R. Felix Geyer and Johannes van der Zouwen (eds.), *Sociocybernetic Paradoxes: Observation, Control and Evolution of Self-Steering Systems* (London: SAGE, 1986), 119–130.

65. Clifford Nass and Scott Brave, *Wired for Speech: How Voice Activates and Advances the Human-Computer Relationship* (Cambridge, MA: MIT Press, 2005), ix.

66. Nass and Brave, *Wired for Speech*, x.

67. Clifford Nass and Li Gong, "Speech Interfaces from an Evolutionary Perspective," *Communications of the ACM*, 43, no. 9 (September 2000): 36–43.

68. Nass and Brave, *Wired for Speech*, 13.

69. Nass and Brave, *Wired for Speech*, 15.

70. Nass and Brave, *Wired for Speech*, 20.

71. Nass and Brave, *Wired for Speech*, 28.

72. Anne Eisenberg, "Mars and Venus, On the Net: Gender Stereotypes Prevail," *New York Times*, October 12, 2000, G1.

73. Henton, "Where Is Female Synthetic Speech?"

74. Lisa Guernsey, "The Desktop That Does Elvis." Other members of the Natural Voices team at Bell Labs included Juergen Schroeter, Alistair Conkie, and Mark Beutnagel.

75. Yolande Strengers and Jenny Kennedy, *The Smart Wife: Why Siri, Alexa, and Other Smart Home Devices Need a Feminist Reboot* (Cambridge, MA: MIT Press, 2021).

76. "Most Downloaded Sat Nav Voice," Guinness World Records, https://www.guinnessworldrecords.com/world-records/most-downloaded-sat-nav-voice.

77. "Morgan Freeman Has the Most Calming Voice of All Time," Netflix: Behind the Stream, YouTube video, 5:32, https://www.youtube.com/watch?v=lwf8rPvLajE. Freeman did, in fact, record a few messages for Zuckerberg's system.

78. "Our Story," VocaliD, https://vocalid.ai/about-us/.

79. VocaliD website, and also Rupal Patel, "Synthetic Voices, as Unique as Fingerprints," TED-Women, TED video, 11:31, December 2013, https://www.ted.com/talks/rupal_patel_synthetic_voices_as_unique_as_fingerprints.

80. Nadine Cortez, "Testing VOCALISE Spoof Detection Capabilities Based on CQT Dilated ResNet Model Between Authentic and Synthetic Voice Samples by VocaliD," *Journal of Emerging Forensic Sciences Research* 6, no. 1 (2021): 47–61.

CHAPTER 6

1. Mark Barton, a cocreator of the voice synthesis software that was being used for the demo, has stated that the scene is "100 percent fiction" and the voice worked perfectly, but the full graphical presentation was too large for the Mac's 128 kilobytes (KB) of memory, so a prototype "fat Mac," with 512 KB of memory, was used. Kay Savetz and Mark Barton, "Software Automatic Mouth: Mark Barton," ANTIC Interview 385, May 22, 2020, audio, 55:23, https://ataripodcast.libsyn.com/antic-interview-385-software-automatic-mouth-mark-barton.

2. Hayley Tsukayama, "Siri Shows off Her Sense of Humor," *Washington Post*, October 12, 2011, https://www.washingtonpost.com/business/technology/siri-shows-off-her-sense-of-humor/2011/10/12/gIQAshfwfL_story.html.

3. "Crazy Ones," directed by Jennifer Golub (for Chiat\Day), 1997, https://www.youtube.com/watch?v=5sMBhDv4sik. The ad won the 1998 Emmy Award for Best Commercial.

4. Folklore has it that someone who attended Homebrew Computer Club meetings had programmed a computer to sing "Daisy Bell," in homage to Bell Labs/HAL.

5. "Computers by the Millions," *SIGPC Newsletter* 5, no 1–2 (Fall-Winter 1982/1983), reposted at Jef Raskin's website, https://www.digibarn.com/friends/jef-raskin/writings/millions.html.

6. Steven Levy, *Insanely Great: The Life and Times of Macintosh, the Computer That Changed Everything* (New York: Penguin, 1994), 122.

7. "Bell System Has New Teaching Aids for High School, Elementary Science," *Journal of the Telephone Industry*, February 27, 1965, 20.

8. Savetz and Barton, "Software Automatic Mouth."

9. John Sculley with John A. Byrne, *Odyssey: Pepsi to Apple . . . A Journey of Adventure, Ideas, and the Future* (New York: HarperCollins, 1987), 403.

10. Sculley and Byrne, *Odyssey*, 403.

11. Jonathan Grudin, "The Computer Reaches Out: The Historical Continuity of Interface Design," *Proceedings of the CHI'89 Conference on Human Factors in Computer Systems*, Seattle, April 14–18, 1990.

12. This metaphor of connection between the computer mind and the human mind is also embedded in the common usability study method of "think aloud" protocols, in which users' goals are identified by having them verbalize what they are doing while using a system or software.

13. Alvy Ray Smith, *A Biography of the Pixel* (Cambridge, MA: MIT Press, 2021).

14. J. C. R. Licklider, "Man-Computer Symbiosis," *IRE Transactions on Human Factors in Electronics*, vol. HFE-1 (March 1960): 4–11; J. C. R. Licklider, "The Computer as a Communication Device," *Science and Technology* 76, no. 2 (April 1968): 1–4.

15. Alan Kay and Adele Goldberg, "Personal Dynamic Media," *Computer* 10, no. 3 (March 1977): 31.

16. Ames, *Charisma Machine*.

17. Kay and Goldberg, "Personal," 32.

18. Kay and Goldberg, "Personal," 32 (emphasis in original).

19. Hugh Dubberly, "The Making of Knowledge Navigator," Dubberly Design Office website, http://www.dubberly.com/articles/the-making-of-knowledge-navigator.html. See also Bill Buxton, *Sketching User Experiences: Getting the Design Right and the Right Design* (New York: Morgan Kaufman, 2007).

20. There were actually different versions of the Knowledge Navigator video, developed for multiple audiences. The EduComm video is the focus of my discussion here.

21. Savetz and Barton, "Software Automatic Mouth."

22. Bianca Bosker, "Siri Rising: The Inside Story of Siri's Origins and Why She Could Over-shadow the iPhone," *Huffington Post*, January 22, 2013, https://www.huffpost.com/entry/siri -do-engine-apple-iphone_n_2499165.

23. Bosker, "Siri Rising."

24. DARPA programs web page, archived on August 5, 2011, at http://www.darpa.mil/Our _Work/I2O/Programs/Personalized_Assistant_that_Learns_(PAL).aspx.

25. Douglas C. Engelbart, "Augmenting Human Intellect: A Conceptual Framework," SRI Summary Report AFOSR-3223, October 1962, archived December 23, 2008 at https://web .archive.org/web/20110504035147/http://www.dougengelbart.org/pubs/augment-3906 .html; see also "Augmenting Society's Collective IQ," Doug Engelbart Institute website, https://dougengelbart.org/content/view/194/.

26. This is documented in Thierry Bardini, *Bootstrapping: Douglas Engelbart, Coevolution, and the Origins of Personal Computing* (Palo Alto, CA: Stanford University Press, 2000).

27. Engelbart cofounded the Doug Engelbart Institute with his daughter Christina Engelbart in 1988 to continue boosting his ideas about Collective IQ.

28. Danielle Newnham, "The Story Behind Siri and the Man Who Made Her," *The Startup*, August 21, 2015, https://medium.com/swlh/the-story-behind-siri-fbeb109938b0.

29. Newnham, "Story Behind Siri."

30. Nuance continued to acquire speech technologies up until 2021, when it was itself acquired by Microsoft for almost $20 billion.

31. Steve Wildstrom, "Nuance Exec on iPhone 4S, Siri, and the Future of Speech," *Tech.pinions*, October 10, 2011, https://techpinions.com/nuance-exec-on-iphone-4s-siri-and-the-future -of-speech/3307.

32. M. G. Siegler, "Siri, Do You Use Nuance Technology? Siri: I'm Sorry, I Can't Answer That," *TechCrunch*, October 5, 2011, https://techcrunch.com/2011/10/05/apple-siri-nuance/.

33. Suchman, "Talk with Machines, Redux," 77.

34. This is not to imply that the extraction of natural resources is not a contemporary problem. Indeed, I recognize it as the root of every environmental problem that humans have created and continue to create. The material infrastructure of information capitalism is itself environmentally catastrophic.

35. Mark Weiser, "The Computer for the 21st Century," *Scientific American*, September 1991, 94–104.

36. Paul Dourish and Genevieve Bell, *Divining a Digital Future: Mess and Mythology in Ubiquitous Computing* (Cambridge, MA: MIT Press, 2014), 2. Dourish and Bell explain mythmaking as "stories that animate individuals and societies by providing paths to transcendence that lift people out of the banality of everyday life. They offer an entrance to another reality; a reality once characterized by the promise of the sublime."

37. Dourish and Bell, *Divining*, 3; Dourish and Bell note that competing research agendas emerged from specific labs in the early 2000s, such as "pervasive computing" at IBM, but all shared a focus on the wide deployment of networked devices.

38. Weiser, "21st Century," 104. See also John Tinnell, *The Philosopher of Palo Alto: Mark Weiser, Xerox PARC, and the Original Internet of Things* (Chicago: University of Chicago Press, 2023).

39. Dourish and Bell, *Divining*, ch. 2. For discussion of previous visions of smart homes, see Lynn Spigel, *Welcome to the Dreamhouse: Popular Media and Postwar Suburbs* (Durham, NC: Duke University Press, 2001); Davin Heckman, *A Small World: Smart Houses and the Dream of the Perfect Day* (Durham, NC: Duke University Press, 2008).

40. Dourish and Bell, *Divining*, 163.

41. Rachel Plotnick, *Power Button: A History of Pleasure, Panic, and the Politics of Pushing* (Cambridge, MA: MIT Press, 2018).

42. Margaret Davidson, "Journal About Home," *Ladies Home Journal*, May 1966, page d.

43. Dag Spicer, "The ECHO IV Home Computer: 50 Years Later," Computer History Museum website, May 31, 2016, https://computerhistory.org/blog/the-echo-iv-home-computer-50-years-later/.

44. Quoted by Spicer, "ECHO IV."

45. *House & Garden*, December 1966, 30; *Family Weekly*, December 1966, 2.

46. Jim Sutherland, "Living with ECHO-IV," presentation in Monroeville, PA, on January 13, 2015, uploaded by the Computer History Museum, February 21, 2018, YouTube video, 49:25, https://www.youtube.com/watch?v=uF1YzWtGRjM.

47. The article quotes "Professor E. L. Kelly" and then refers to this person using the female pronoun "her." I wonder if this wasn't a typo, and that "E. L. Kelly" is the noted psychologist Everett Lowell Kelly, who studied marital compatibility in the World War II era. Kelly's survey instrument bears the marks of patriarchal midcentury gender role assumptions. See E. Lowell Kelly, "Marital Compatibility as Related to Personality Traits of Husbands and Wives as Rated by Self and Spouse," *Journal of Social Psychology* 13, no. 1 (February 1, 1941): 193–198.

48. David Bill Hempstead, "What the 'Think Machines' Will Do for You," *Redbook*, May 1955, 46–47.

49. Hempstead, "What the 'Think Machines' Will Do," 47.

50. Daniela Hernandez, "Before the iPad, There Was the Honeywell Kitchen Computer," *Wired*, November 22, 2012, https://www.wired.com/2012/11/kitchen-computer/.

51. Richard Fry, Carolina Aragão, Kiley Hurst, and Kim Parker, "In a Growing Share of U.S. Marriages, Husbands and Wives Earn About the Same," Pew Research Center report, April 13, 2023, https://www.pewresearch.org/social-trends/2023/04/13/in-a-growing-share-of-u-s-marriages-husbands-and-wives-earn-about-the-same/.

52. Alan P. Hald, "Toward the Information-Rich Society," *Futurist*, August 1981, 20–24.

53. Hald, "Toward the Information-Rich Society," 20.

54. Hald, "Toward the Information-Rich Society," 21.

55. John Blankenship, *The Apple House: How to Computerize Your Home Using Your Apple Computer* (Hoboken, NJ: Prentice-Hall, 1984), vi.

56. Blankenship, *Apple House*, 1.

57. Blankenship, *Apple House*, 2.

58. Blankenship, *Apple House*, 27.

59. In their study of acousmatic computer voices in movies, Liz Faber says that the combination of the British manservant archetype, effeminate and intellectual, together with Stark's American rugged masculinity, combine in the Iron Man suit in a symbiotic relationship of brains and brawn. Faber, *Computer's Voice*.

60. *Smart House*, directed by LeVar Burton (Los Angeles: Alan Sacks Productions, 1999).

61. Arielle Pardis, "The AI-Fueled, Anxious Hopefulness of Disney's *Smart House*," *Wired*, June 26, 2019, https://www.wired.com/story/disney-channel-smart-house-20-years-later/.

62. Christoph Bartneck, Michel Van Der Hoek, Omar Mubin, and Abdullah Al Mahmud, "Daisy, Daisy, Give Me Your Answer Do! Switching off a Robot," in *Proceedings of the ACM/IEEE International Conference on Human-Robot Interaction* (Washington, DC: ACM, 2007), 219.

63. Halcyon M. Lawrence, "Siri Disciplines," in Thomas S. Mullaney, Benjamin Peters, Mar Hicks, and Kavita Philip (eds.), *Your Computer Is On Fire* (Cambridge, MA: MIT Press, 2021), 179–197.

64. Joseph Turow, *The Voice Catchers: How Marketers Listen In to Exploit Your Feelings, Your Privacy, and Your Wallet* (New Haven, CT: Yale University Press, 2021). On accent detection, see also Belle Lin, "Amazon's Accent Recognition Technology Could Tell the Government Where You're From," *The Intercept*, November 15, 2018, https://theintercept.com/2018/11/15/amazon-echo-voice-recognition-accents-alexa/.

65. Jack Gillum and Jeff Kao, "Aggression Detectors: The Unproven, Invasive Surveillance Technology Schools Are Using to Monitor Students," *ProPublica*, June 25, 2019, https://features.propublica.org/aggression-detector/the-unproven-invasive-surveillance-technology-schools-are-using-to-monitor-students/.

66. Richard Weissbourd, Milena Batanova, Virginia Lovison, and Eric Torres, "Loneliness in America: How the Pandemic Has Deepened an Epidemic of Loneliness and What We Can Do About It," preliminary report, Harvard Graduate School of Education, February 2021, https://mcc.gse.harvard.edu/reports/loneliness-in-america.

67. Sherry Turkle, *Alone Together* (New York: Basic Books, 2011); and *Reclaiming Conversation: The Power of Talk in a Digital Age* (New York: Penguin Press, 2015).

68. *Operator*, directed by Logan Kibens (Chicago: Cruze & Company, 2018).

EPILOGUE

1. Frank Rooney, "A Friend of Charlie's," *Cosmopolitan*, July 1950, 117.

2. Jo-Anne Bachorowski, Moria J. Smoski, and Michael J. Owren, "The Acoustic Features of Human Laughter," *Journal of the Acoustical Society of America* 110, no. 3 (2001): 1581.

3. Roger Ebert, "Remaking My Voice," TED video, 19:13, March 2011, https://www.ted.com /talks/roger_ebert_remaking_my_voice.

4. Ebert, "Remaking My Voice."

5. Joke sample, Cerewave AI, CereProc website, https://www.cereproc.com/en/v6.

6. "JFK Unsilenced," CereProc website, https://www.cereproc.com/en/jfkunsilenced.

7. Suchman, "Talk with Machines, Redux," 78.

8. Judith Newman, "To Siri, with Love," *New York Times*, October 17, 2014, https://www .nytimes.com/2014/10/19/fashion/how-apples-siri-became-one-autistic-boys-bff.html; print, New York edition, October 19, 2014, Section ST, 1. A book-length expansion of this article, published with the same title, received a lot of criticism and calls for boycotts from some in the autistic community. Judith Newman, *To Siri with Love: A Mother, Her Autistic Son, and the Kindness of Machines* (New York: Harper, 2017).

9. Kenneth Olmstead, "Nearly Half of Americans Use Digital Assistants Mostly on Their Smartphones," Pew Research Center, December 12, 2017, https://www.pewresearch.org /fact-tank/2017/12/12/nearly-half-of-americans-use-digital-voice-assistants-mostly-on-their -smartphones/.

10. Grayson Kemper, "Virtual Assistants and Consumer AI," Clutch Report, February 11, 2019, https://clutch.co/bpo/virtual-assistant/resources/virtual-assistants-consumer-ai.

11. In June 2023, Senator Jon Ossoff (D-GA), the chair of the Senate Human Rights Subcommittee, convened a bipartisan hearing to explore the implications of AI for human rights. One woman, Jennifer DeStefano of Arizona, testified about her experience receiving a phone call from people who used a synthesized deep fake of her teenaged daughter's voice to try to convince DeStefano that her daughter had been kidnapped and held for ransom. See https:// www.ossoff.senate.gov/press-releases/watch-sen-ossoff-convenes-hearing-on-implications-of -artificial-intelligence-for-human-rights/.

Index

Bionics, 137–138
British Broadcasting Corporation (BBC),
    55–56
Bush, Vannevar, 25–26, 50, 92, 140, 178

Čapek, Karel, 46–47, 48. See also *Rossum's*
    Universal Robots (R.U.R.)
Chaplin, Charlie. See *Modern Times*
    (film)
Cold War, 68, 74–75, 96, 104, 128, 178,
    184. *See also* Space race
Comedy, 22, 67, 190
    about computers, 67–69, 71–73, 97–98
    stand-up, 69
Computerization, 77, 98, 119. *See also*
    Automation; Microprocessors
    of financial institutions, 87–94, 152
    of households, 179–184
    of manufacturing, 134
    of office work, 74
Computers. *See also* Apple; International
    Business Machines; UNIVAC
    anthropomorphizing of, 51–52, 60, 66,
        77, 92–93, 131–133
    artistic creativity with, 75–76, 133 (*see also*
        Experiments in Art and Technology)
    brain metaphor, 5, 8, 10, 51, 57–60, 66,
        68, 73, 76, 100–101, 129, 131–132,
        135, 138, 140 (*see also* Artificial
        intelligence; Cybernetics)
    chips, 123–126, 131, 154 (*see also*
        Integrated circuits; Microprocessors)
    digital, 78, 141
    educational, 108, 119, 128, 134, 138
        (*see also* Educational technology)
    fictional, 105, 118, 133–136, 152–153,
        186, 195–196
    home, 78, 108, 119, 134, 142, 164,
        180–181
    interfaces for, 7, 92, 145, 156, 164,
        173–175 (*see also* Voice synthesis)
    mainframe, 52, 58, 68, 77, 97, 134

    predicting the future, 51, 61–63, 66, 70,
        72–73, 76, 89, 102, 195
    social relationships with, 78, 97–98,
        130–133, 135–136, 156, 165, 187–188
    talking, 104–105, 118, 133–136, 154,
        169, 187
    on television, 52, 66, 69, 97–99, 133–136
    user friendly, 87, 163, 165–166, 168, 172,
        176, 190
Computer Wore Tennis Shoes, The, 98–104,
    179, 186
Consumer electronics, 87, 107, 111
    for children and young people, 108,
        118–133
    marketing, 109, 125
    Merlin (game), 122–126, 129
    Simon (game), 122, 125–126, 129–130
    talking, 108, 110, 127, 139–140, 143,
        145, 154, 160, 165
Consumerism, 43–44, 48, 84, 98, 181
Cybernetics, 5, 15, 59, 76, 88, 94, 96, 100,
    122, 135, 156, 168–169
Cyborg, 137–138, 147, 151, 161, 226n1

Data determinism, 66
DEC. *See* Digital Equipment Corporation
Descartes, René, 2, 15
Digital Equipment Corporation (company),
    139
    DECTalk, 147, 167
    PDP-9 (computer), 143
Dodge, Charles, 84–85. See also *Speech Songs*
    (album)
Dudley, Homer W., 29–34, 36–37, 52–53,
    57, 86, 113, 140
Dunn, H. K., 53–55, 57

Eames, Charles, 75–76, 105, 127, 133,
    165, 183. See also *Information*
    Machine, The
Educational technology, 107, 109, 116–117,
    119, 126–131, 138, 143, 170